Second Siberian Winter School

Algebra and Analysis

Recent Titles in This Series

151 **I. A. Aleksandrov, L. A. Bokut′, and Yu. G. Reshetnyak, Editors,** Second Siberian Winter School "Algebra and Analysis"
150 **S. G. Gindikin, Editor,** Spectral Theory of Operators
149 **V. S. Afraĭmovich, et al.,** Thirteen Papers in Algebra, Functional Analysis, Topology, and Probability, Translated from the Russian
148 **A. D. Aleksandrov, O. V. Belegradek, L. A. Bokut′, and Yu. L. Ershov, Editors,** First Siberian Winter School in Algebra and Analysis
147 **I. G. Bashmakova, et al.,** Nine Papers from the International Congress of Mathematicians 1986
146 **L. A. Aĭzenberg, et al.,** Fifteen Papers in Complex Analysis
145 **S. G. Dalalyan, et al.,** Eight Papers Translated from the Russian
144 **S. D. Berman, et al.,** Thirteen Papers Translated from the Russian
143 **V. A. Belonogov, et al.,** Eight Papers Translated from the Russian
142 **M. B. Abalovich, et al.,** Ten Papers Translated from the Russian
141 **Kh. Drashkovicheva, et al.,** Ordered Sets and Lattices
140 **V. I. Bernik, et al.,** Eleven Papers Translated from the Russian
139 **A. Ya. Aĭzenshtat, et al.,** Nineteen Papers on Algebraic Semigroups
138 **I. V. Kovalishina and V. P. Potapov,** Seven Papers Translated from the Russian
137 **V. I. Arnol′d, et al.,** Fourteen Papers Translated from the Russian
136 **L. A. Aksent′ev, et al.,** Fourteen Papers Translated from the Russian
135 **S. N. Artemov, et al.,** Six Papers in Logic
134 **A. Ya. Aĭzenshtat, et al.,** Fourteen Papers Translated from the Russian
133 **R. R. Suncheleev, et al.,** Thirteen Papers in Analysis
132 **I. G. Dmitriev, et al.,** Thirteen Papers in Algebra
131 **V. A. Zmorovich, et al.,** Ten Papers in Analysis
130 **M. M. Lavrent′ev, et al.,** One-dimensional Inverse Problems of Mathematical Physics
129 **S. Ya. Khavinson; translated by D. Khavinson,** Two Papers on Extremal Problems in Complex Analysis
128 **I. K. Zhuk, et al.,** Thirteen Papers in Algebra and Number Theory
127 **P. L. Shabalin, et al.,** Eleven Papers in Analysis
126 **S. A. Akhmedov, et al.,** Eleven Papers on Differential Equations
125 **D. V. Anosov, et al.,** Seven Papers in Applied Mathematics
124 **B. P. Allakhverdiev, et al.,** Fifteen Papers on Functional Analysis
123 **V. G. Maz′ya, et al.,** Elliptic Boundary Value Problems
122 **N. U. Arakelyan, et al.,** Ten Papers on Complex Analysis
121 **D. L. Johnson,** The Kourovka Notebook: Unsolved Problems in Group Theory
120 **M. G. Kreĭn and V. A. Jakubovič,** Four Papers on Ordinary Differential Equations
119 **V. A. Dem′janenko, et al.,** Twelve Papers in Algebra
118 **Ju. V. Egorov, et al.,** Sixteen Papers on Differential Equations
117 **S. V. Bočkarev, et al.,** Eight Lectures Delivered at the International Congress of Mathematicians in Helsinki, 1978
116 **A. G. Kušnirenko, A. B. Katok, and V. M. Alekseev,** Three Papers on Dynamical Systems
115 **I. S. Belov, et al.,** Twelve Papers in Analysis
114 **M. Š. Birman and M. Z. Solomjak,** Quantitative Analysis in Sobolev Imbedding Theorems and Applications to Spectral Theory
113 **A. F. Lavrik,** Twelve Papers in Logic and Algebra
112 **D. A. Gudkov and G. A. Utkin,** Nine Papers on Hilbert's 16th Problem

(*Continued in the back of this publication*)

American Mathematical Society

TRANSLATIONS

Series 2 • Volume 151

Soviet Regional Conferences

Second Siberian Winter School

Algebra and Analysis

Proceedings of the Second Siberian School
Tomsk State University, Tomsk
1989

I. A. Aleksandrov
L. A. Bokut′
Yu. G. Reshetnyak
Editors

American Mathematical Society
Providence, Rhode Island

Translation edited by SIMEON IVANOV

1991 *Mathematics Subject Classification.* Primary 11Sxx, 14F05, 17-xx, 32-xx, 46S20; Secondary 11Fxx, 20F32, 30-xx, 58-xx.

Library of Congress Cataloging-in-Publication Data
Siberian Winter School "Algebra and Analysis" (2nd: 1989: Tomsk, R.S.F.S.R.)
[Algebra i analiz. English]
Second Siberian Winter School "Algebra and Analysis": proceedings of the Second Siberian School, Tomsk State University, Tomsk, 1989/ I. A. Aleksandrov, L. A. Bokut, Yu. G. Reshetnyak, editors.
p. cm. - (American Mathematical Society translations, ISSN 0065-9290; ser. 2, v. 151. Soviet regional conferences)
Translation of: Algebra i analiz.
ISBN 0-8218-3142-9 (alk. paper)
1. Algebra—Congresses. 2. Mathematical analysis—Congresses. I. Bokut, L. A. (Leonid A.), 1937- . II. Title. III. Series: American Mathematical Society translations; ser. 2, v. 151. IV. Series: American Mathematical Society translations. Soviet regional conferences.
QA3.A572 Ser. 2, vol. 151
[QA150]
510 s–dc20 92-15960
[512′.15] CIP

The paper used in this book is acid-free and falls within the guidelines
established to ensure permanence and durability. ♾

This publication was typeset using AMS-TEX,
the American Mathematical Society's TEX macro system.

10 9 8 7 6 5 4 3 2 1 97 96 95 94 93 92

Contents

Amer. Math. Soc. Transl.
(2) Vol. **151**, 1992

On Gradings of Simple Lie Algebras Connected with Groups Generated by Transvections

UDC 512.81

A. V. ALEKSEEVSKIĬ

The basis of Cartan theory is a one-to-one correspondence between semisimple Lie algebras and integral groups generated by reflections (Weyl groups). To an abstract Weyl group W corresponds a semisimple Lie algebra $\mathfrak{J}$ and a decomposition $\mathfrak{J} = \mathfrak{J}_0 \dot{+} \Sigma^{\boldsymbol{\cdot}} \mathfrak{J}_\alpha$ of $\mathfrak{J}$ into a direct sum of root subspaces $\mathfrak{J}_\alpha$ and a Cartan subalgebra $\mathfrak{J}_0$. The root decomposition is a grading, i.e., $[\mathfrak{J}_\alpha, \mathfrak{J}_\beta] \subset \mathfrak{J}_{\alpha+\beta}$, that cannot be refined (a grading $\mathfrak{J} = \Sigma^{\boldsymbol{\cdot}} \mathfrak{J}_\omega$ is called a refinement of a grading $\mathfrak{J} = \Sigma^{\boldsymbol{\cdot}} \mathfrak{J}_\alpha$ if for each α there exist $\omega_1, \dots, \omega_k$ such that $\mathfrak{J}_\alpha = \mathfrak{J}_{\omega_1} \dot{+} \cdots \dot{+} \mathfrak{J}_{\omega_k}$).

It is known that the gradings of simple Lie algebras, even the unrefinable ones, are not exhausted by the root decompositions. A simple example can be constructed from the three Pauli matrices $\sigma_1, \sigma_2, \sigma_3$ constituting a basis of the Lie algebra SL_2. The commutator of any two of them is proportional to the third. Therefore the decomposition of SL_2 into the sum of the three one-dimensional subspaces spanned by the Pauli matrices is a grading relative to the finite Abelian group $\mathbb{Z}_2 \times \mathbb{Z}_2$, with the subspaces $\mathfrak{J}_0$ corresponding to the identity element of the group being the zero subspace.

In the present paper, a correspondence similar to the one between Weyl groups and root decompositions is established between other objects: a class of gradings of simple Lie algebras, on the one hand, and linear groups over a finite prime field $\mathbb{F}_p$ that are generated by transvections ([1]) (instead of Weyl groups), on the other. The analogue of a Cartan subgroup, in this situation, is some finite commutative subgroup of the group of automorphisms of a Lie algebra, called a Jordan subgroup.

1991 *Mathematics Subject Classification*. Primary 17B20, 17B70, 22E10.

([1]) A transvection is a unipotent linear transformation r such that $\operatorname{rank}(r - e) = 1$ (e is the identity transformation).

In order to explain how this correspondence is constructed we turn first to Cartan theory (see, e.g., [5]).

1. From a semisimple algebraic group G and a Cartan subgroup (maximal torus) T there is constructed an integral linear group $W = W(G, T)$ (the Weyl group).

The Weyl group W is defined as the factor group of the normalizer $N_G(T)$ of the maximal torus with respect to the centralizer $Z_G(T)$. The group W acts naturally as automorphisms of the integral lattice of characters $\mathfrak{X} = \mathrm{Hom}(T, k^*)$ of the torus T (k is the ground field, which, for definiteness, we take to be algebraically closed of characteristic zero).

The theorem asserting that the Weyl group is generated by reflections can be proved as follows. One constructs a root decomposition of the Lie algebra $\mathfrak{J}$ of G, i.e., a decomposition $\mathfrak{J} = \mathfrak{J}_0 + \Sigma^{\cdot} \mathfrak{J}_\alpha$ of $\mathfrak{J}$ into a sum of weight subspaces relative to the action of a maximal torus by adjoint operators. The weights of T occurring in this decomposition are called roots. The set $\Omega = \Omega(G, T)$ of these roots is called a root system of G. With each line spanned by a root α there is associated a root subgroup $G_{(\alpha)}$ of G. Its Lie algebra is generated by the subspaces of the form $\mathfrak{J}_{\lambda\alpha}$ (in fact, $\lambda = \pm 1$). It can be proved that all root subgroups are isomorphic to SL_2. In each subgroup $G_{(\alpha)}$ there is an element $\bar{r}(\alpha)$ normalizing the torus T. Under the epimorphism $N_G(T) \to W$ this element passes into a reflection $r_{(\alpha)}$. It can be proved that the group W is generated by all $r_{(\alpha)}$.

2. There is given an axiomatic definition of an abstract Weyl group. With each such group there is associated an abstract root system. It can be proved that an abstract Weyl group is completely determined by its root system. All root systems and, therefore, all abstract Weyl groups can be classified.

3. The following existence theorem can be proved (by a direct construction of a Lie algebra in a root decomposition with respect to a root system): If W is any abstract Weyl group, then there exist a semisimple algebraic group G and a maximal torus T such that the Weyl group $W(G, T)$ is isomorphic to W as a linear group.

4. The following uniqueness theorem can be proved: Two Weyl groups $W(G, T)$ and $W(G', T')$ of the groups G and G' are isomorphic as linear groups if and only if there exists an isomorphism $\varphi: G \simeq G'$ sending the Cartan subgroup T of G into the Cartan subgroup T' of G'.

5. It can be proved that any two Cartan subgroups T, S of a semisimple algebraic group G are conjugate in G.

We intentionally left the conjugacy theorem for Cartan subgroups to the last step. It is easy to show that the entire theory can be developed independently of this theorem, which is usually proved first. We did this because, as it happens, steps 1–4 can be carried out for certain other diagonalizable subgroups of G in place of the Cartan subgroups (or, equivalently, for certain gradings of the Lie algebra $\mathfrak{J}$ different from the root decompositions).

The role of a Cartan subgroup in this paper is played by a Jordan subgroup j of a simple algebraic group G. A Jordan subgroup is isomorphic to a product of simple cyclic groups $\mathbb{Z}_p$ and is not contained in any torus.

Exactly as in the Cartan theory we can define a linear group $W = W(G, j)$, which we call the Weyl J-group. It acts in a linear space over the field $\mathbb{F}_p$, since that is how we can view the dual group $J = \mathrm{Hom}(j, k^*)$. In this paper we will prove that the group W is generated by transvections (the transvections are taken from the root subgroups, which in this case are tori).

With a Jordan subgroup j is associated a bilinear form preserved by the Weyl J-group. This form is skew-symmetric and, generally speaking, can be zero. Jordan subgroups for which this form is nondegenerate exist only in classical simple algebraic groups. Therefore such Jordan subgroups will be called classical Jordan subgroups.

It can be proved that the simple groups with a distinguished classical Jordan subgroup are in one-to-one correspondence with the irreducible linear groups over $\mathbb{F}_p$ generated by transvections and preserving a nondegenerate bilinear form. From this fact and the well-known classification of linear groups generated by transvections (see [2], [9]) we can obtain a classification of classical Jordan subgroups. All nonclassical Jordan subgroups can also be found in this paper.

We should mention that Jordan subgroups do not exist in all simple algebraic groups, and for such groups, generally speaking, the conjugacy theorem (the fifth step) is not true.

Since with each Jordan subgroup there is associated a grading of the simple Lie algebra (the root decomposition relative to the Jordan subgroup), we obtain a description of all such gradings. They are of independent interest. Those associated with classical Jordan subgroups were studied earlier by Popovici [11]. They can be used to construct, in a uniform way, a basis of a simple Lie algebra with respect to which the structure constants have a particularly simple form. Thus we can obtain, for example, a basis of the Lie algebra of the orthogonal group connected with the Clifford algebra.

A complete list of all gradings connected with nonclassical Jordan subgroups was given in [1]. The grading of the Lie algebra E_8 associated with one of the two Jordan subgroups of E_8 was constructed independently by Thompson [15] and was used to construct a simple finite group in [13]. The same grading of E_8 and a grading of the Lie algebra G_2 associated with a Jordan subgroup were described by Hesselink [18] among all gradings of Lie algebras in which each subspace $\mathfrak{J}_\alpha$ is a Cartan subalgebra. In developing the methods of [15], [13] with the aim of realizing simple finite groups as groups of automorphisms of lattices lying in a Lie algebra, Kostrikin and his coauthors introduced the concept of an orthogonal Cartan decomposition of a simple Lie algebra [19]. It is noteworthy, in our opinion, that many of the constructed examples of orthogonal Cartan decompositions ([19], [20], [21], [22]) can be obtained by summing root subspaces of gradings associated with Jordan subgroups.

Jordan subgroups are interesting in their own right. They appear in the study of maximal solvable subgroups. Similar subgroups of classical linear groups over finite fields were first considered by Jordan in his papers concerned with describing all maximal solvable subgroups of the symmetric group and some other finite groups (see [17]). Jordan subgroups of the full linear group over an arbitrary field were considered by Suprunenko [6] in connection with the same problem.

The term "Jordan subgroup" was introduced by this author in [1]. This concept was needed to generalize some of Suprunenko's results on maximal solvable subgroups to other simple algebraic groups [23] and to study the maximal finite subgroups of these groups [24]. Also stated in [1] were the main theorems on the connections among Jordan subgroups, gradings of simple Lie algebras, and groups generated by transvections. Here we will take advantage of the opportunity to publish complete proofs.

All of our main concepts and theorems are stated in §1. In §2 we mention without proof some necessary facts from the theory of groups generated by transvections. In §3 and §4 we prove all of the main theorems (corresponding to the first, third, and fourth steps). In §5 we describe the nonclassical Jordan subgroups.

§1. Statement of the main results

We will denote by G a connected simple algebraic group, defined over the field of complex numbers $\mathbb{C}$. ([2]) As a rule, we will assume that the center of G is trivial, which is equivalent to saying that the adjoint representation Ad_G is faithful. We will define the concept of a Jordan subgroup of G.

DEFINITION 1.1. A diagonalizable subgroup (i.e., a commutative subgroup consisting of semisimple elements) j of G is called a Jordan subgroup if the following conditions are satisfied:

(1) (Irreducibility of the normalizer) j is a minimal normal subgroup of its normalizer $N_G(j)$.

(2) (Nonsimplification of the normalizer) If H is a connected reductive subgroup of G that is invariant under conjugation by all elements of $N_G(j)$, then H is a simple group and its centralizer in G is trivial.

(3) (Maximality of the normalizer) If k is a diagonalizable subgroup of G such that $N_G(k) \supset N_G(j)$, then $N_G(k) = N_G(j)$.

The definition of a Jordan subgroup is very technical. For one motivation of the definition see [23], [24]. Within the scope of the present paper,

([2]) All of our results are true verbatim for an algebraically closed field of characteristic 0.

however, justification of the definition is provided by the theorems stated below.

It follows easily from condition (2) of the definition that a Jordan subgroup j and its normalizer $N_G(j)$ are finite groups.

We give an example of a Jordan subgroup.

PROPOSITION 1.2. *Suppose $G = \mathrm{PSL}_n(\mathbb{C})$ and $n = p^\alpha$, where p is a prime. Then G contains a Jordan subgroup $j \simeq (\mathbb{Z}_p \times \mathbb{Z}_p)^\alpha$. It is a tensor product of subgroups $j_i \simeq \mathbb{Z}_p \times \mathbb{Z}_p$ lying in $\mathrm{PSL}_p(\mathbb{C})$:*

$$j = j_1 \times \cdots \times j_\alpha \subset p(\mathrm{SL}_p(\mathbb{C}) \otimes \cdots \otimes \mathrm{SL}_p(\mathbb{C})).$$

Each of the subgroups j_i is generated by two transformations $a, b \in \mathrm{SL}_p(\mathbb{C})$ with matrices

$$a = \begin{pmatrix} 1 & & & \\ & \varepsilon & & \\ & & \ddots & \\ & & & \varepsilon^{p-1} \end{pmatrix}, \quad \varepsilon^p = 1; \qquad b = \begin{pmatrix} 0 & \cdots & 0 & 1 \\ 1 & 0 & \cdots & 0 \\ 0 & 1 & 0 & \cdots \\ \cdots & \cdots & \cdots & \cdots \end{pmatrix}.$$

The factor group $N_G(j)/j$ is isomorphic to the group of symplectic transformations over the field of p elements: $N_G(j)/j \simeq \mathrm{Sp}_{2\alpha}(\mathbb{F}_p)$.

Proposition 1.2 follows from the results of Suprunenko [6], where these subgroups are studied in detail.

We introduce the main objects associated with a Jordan subgroup. They are analogous to the Lie algebra $\mathscr{L}$ of a Cartan subgroup T, the Weyl group in the representation in the space $\mathscr{L}$, a root decomposition of a Lie algebra, and a root system in the Cartan theory.

We will first construct a vector space J associated with a Jordan subgroup j, in the same way as the Lie algebra $\mathscr{L}$ of a Cartan subgroup T is associated with the Cartan subgroup itself. It is known that a minimal commutative normal subgroup of a finite group is isomorphic to a direct product of cyclic subgroups of prime order p. It therefore follows from condition (1) of Definition 1.1 that $j \simeq \mathbb{Z}_p \times \cdots \times \mathbb{Z}_p$ ($\mathbb{Z}_p$ is a cyclic subgroup of prime order p).

It is clear that the group $j \simeq \mathbb{Z}_p \times \cdots \times \mathbb{Z}_p$ in additive notation can be viewed as a vector space over the field $\mathbb{F}_p$ of p elements. The space J will be isomorphic to this vector space, but the isomorphisms of the additive group of J and the group j of the dual objects must be chosen in a certain way. Namely, let $j^\#$ be the group of characters of the group j, i.e., $j^\# = \mathrm{Hom}(j, \mathbb{C}^*)$. Let $J = \mathrm{Hom}(j^\#, \mathbb{F}_p^+)$ (as usual, k^* denotes the multiplicative, and k^+ the additive, group of the field k). We choose $\varepsilon = \sqrt[p]{1}$ to be a primitive root and construct an embedding $\mathbb{F}_p \subset \mathbb{C}^*$ by the rule $x \mapsto \varepsilon^x$. By duality, $j = \mathrm{Hom}(j^\#, \mathbb{C}^*)$. This embedding defines an isomorphism $\varepsilon\colon J = \mathrm{Hom}(j^\#, \mathbb{F}_p) \to j = \mathrm{Hom}(j^\#, \mathbb{C}^*)$. The isomorphism of the groups $j^{\#}$ and $J^* = \operatorname{Hom}(J, \mathbb{F}_p)$ that is compatible with ε is denoted by d, $d\colon j^{\#} \simeq J^*$. Compatibility is defined by the formula

$$\varepsilon^{d\lambda(x)} = \lambda(\varepsilon(x)), \qquad x \in J, \ \lambda \in j^{\#}.$$

Thus from a Jordan subgroup j we have constructed a vector space J over $\mathbb{F}_p$. The dimension of J is equal to the rank of the Abelian group j: $\dim_{\mathbb{F}_p}(J) = \operatorname{rank} j$.

We will now construct a representation of the normalizer $N = N_G(j)$ of the Jordan subgroup j in the space J. If $g \in N$, then g acts by conjugation on j: $a \mapsto gag^{-1}$, $a \in j$. The action of g on j can be carried over via the isomorphism $\varepsilon^{-1}\colon j \simeq J$ to a linear action of g in the vector space J, which we denote by $A(g)$, $A(g)\colon J \to J$. It is clear that the constructed mapping $A\colon N \to \mathrm{GL}(J)$ is a representation of N. It follows from the construction of the representation A that the kernel $\operatorname{Ker} A$ of the representation is the centralizer $Z_G(j)$ of the Jordan subgroup j: $\operatorname{Ker} A = Z_G(j)$. The image of the representation A is denoted by W: $W = A(N) = N_G(j)/Z_G(j)$. The linear group W is an analogue of the Weyl group.

DEFINITION 1.3. Suppose j is a Jordan subgroup of G. The group $W = N_G(j)/Z_G(j)$, considered in its natural representation $A\colon W \to \mathrm{GL}(J)$ in the vector space J associated with j, will be called the Weyl J-group of G relative to j.

A central place in our theory is occupied by

THEOREM 1.4. *The Weyl J-group $W = W(G, j)$ of G relative to the Jordan subgroup j is an irreducible linear group over the field $\mathbb{F}_p$ and is generated by transvections.*

Theorem 1.4 will be proven in §3. As in the Cartan theory, an important role in the proof of this theorem is played by the concept of "root system" and root groups. The analogue of a root system, like the classical root systems, can be defined in two different ways.

First, with each linear group W generated by transvections, as with a group generated by reflections, it is natural to associate a set Ω_T of linear forms (the set of axes of the transvections), called the T-root system of W (see §2, Definition 2.2).

Second, with each Jordan subgroup j of G (again as with a Cartan subgroup) there is associated the set of characters of j occurring with nonzero multiplicity in its representation by adjoint operators in the Lie algebra $\mathfrak{J}$ of G. Using the isomorphism $d\colon j^{\#} \simeq J^*$, we can identify the characters of j with linear forms on the vector space J associated with the Jordan subgroup j. The resulting set of linear forms is called a J-root system.

DEFINITION 1.5. Suppose j is a Jordan subgroup of G and J is the vector space over $\mathbb{F}_p$ associated with j. A linear form $\alpha \in J^*$ will be called a J-root of G relative to j if the character ε^{α} of G occurs with nonzero multiplicity in the representation of j by adjoint operators in the Lie algebra $\mathfrak{J}$ of G. The set Ω of all J-roots will be called the J-root system of G relative to j.

An important refinement of Theorem 1.4 is that the two indicated definitions of "root system" are equivalent.

PROPOSITION 1.6. *Suppose j is a Jordan subgroup of G, J is the vector space associated with j, and $W = W(G, j) \subset \mathrm{GL}(J)$ is the Weyl J-group. Then the T-root system $\Omega_T \subset J^*$ of W is the same as the J-root system $\Omega \subset J^*$ of G relative to j, i.e., $\Omega_T = \Omega$.*

Exactly as in the Cartan theory, we can define a root decomposition of the Lie algebra $\mathfrak{J}$ of G. Indeed, since a Jordan subgroup j is diagonalizable (by definition), its representation by adjoint operators in the Lie algebra $\mathfrak{J}$ of G splits into a direct sum $\mathfrak{J} = \dot{\Sigma}\mathfrak{J}_\lambda$ of weight subspaces $\mathfrak{J}_\lambda$, $\lambda \in j^\#$. By definition, $\mathfrak{J}_\lambda \neq \{0\}$ if and only if $d\lambda$ is a J-root.

DEFINITION 1.7. The decomposition $\mathfrak{J} = \dot{\Sigma}_{\lambda\in\Omega}\mathfrak{J}_\lambda$ of the Lie algebra $\mathfrak{J}$ of G with J-root system Ω relative to the Jordan subgroup j is called the root decomposition relative to j.

Below, for convenience, we will parametrize the root subspaces $\mathfrak{J}_\lambda$ by the J-roots $\alpha \in \Omega$, i.e., we will use $\mathfrak{J}_\lambda = \mathfrak{J}_{d^{-1}(\alpha)}$, where $\alpha \in J^*$.

It is clear that the commutator $[\mathfrak{J}_\alpha, \mathfrak{J}_\beta]$ of the root subspaces $\mathfrak{J}_\alpha, \mathfrak{J}_\beta$ $(\alpha, \beta \in \Omega)$ is contained in $\mathfrak{J}_{\alpha+\beta}$. Moreover, $\mathfrak{J}_0$ coincides with the Lie algebra of the centralizer $Z_G(j)$ of the Jordan subgroup j. Since $Z_G(j)$ is a finite group, $\mathfrak{J}_0 = \{0\}$.

LEMMA 1.8. *The root decomposition $\mathfrak{J} = \dot{\Sigma}_{\alpha\in\Omega}\mathfrak{J}_\alpha$ is a grading of the Lie algebra $\mathfrak{J}$ of G via the Abelian group J^*. The zero subspace $\mathfrak{J}_0$ of this grading is trivial: $\mathfrak{J}_0 = \{0\}$.*

Thus from a Jordan subgroup j of G we have constructed the Weyl J-group W, which is a group generated by transvections, and a J-root system Ω.

We will give an axiomatic definition of a Weyl J-group.

DEFINITION 1.9. An irreducible linear group W over a prime field $\mathbb{F}_p$ that is generated by transvections will be called an abstract Weyl J-group.

In the theory of groups generated by transvections it is natural to distinguish the case where the group preserves a nondegenerate bilinear form. This case is of particular interest to us, so we introduce

DEFINITION 1.10. A Jordan subgroup j of G will be called a classical Jordan subgroup if the Weyl J-group $W = W(G, j)$ preserves a nondegenerate bilinear form $c(x, y)$ on the vector space J associated with j. In this case, the group W will be called a Weyl J-group of classical type.

In §3 the bilinear form $c(x, y)$ will be constructed explicitly from the group G and Jordan subgroup j. Note that its nature is completely different from that of the analogous Killing form in the Cartan theory.

It turns out that, as in the Cartan theory, the J-root system is a complete system of invariants of an abstract Weyl J-group of classical type. Abstract Weyl J-groups and, in general, groups generated by transvections have been studied in detail and classified ([8], [9]). Their classification in the case of interest to us is actually based on the classification of T-root systems.

The next step in constructing the theory of Jordan subgroups is to prove an existence theorem.

THEOREM 1.11. *Suppose W is an abstract Weyl J-group of classical type. Then there exist a simple complex algebraic group G and a classical Jordan subgroup j such that the Weyl J-group $W(G, j)$ is isomorphic to W as a linear group*: $W(G, j) \simeq W$.

Theorem 1.11 is proved in §4. The proof consists of an explicit construction of the structure constants of a Lie algebra using a root decomposition.

Finally, as in the Cartan theory, we prove a theorem on the uniqueness of a group G with a classical Jordan subgroup having a prescribed Weyl J-group.

THEOREM 1.12. *Suppose j and j' are classical Jordan subgroups of G and G'. If the Weyl J-groups $W(G, j)$ and $W(G', j')$ are isomorphic as linear groups, then there exists an isomorphism Θ: $G \simeq G'$ of G and G' such that $\Theta(j) = j'$.*

The proof of Theorem 1.12 is given in §4.

Thus we have established a one-to-one correspondence between abstract Weyl J-groups and pairs $(G \supset j)$, where j is a classical Jordan subgroup of a simple complex algebraic group G with trivial center.

As a direct consequence of the theory of Jordan subgroups outlined above and the known classification of abstract Weyl J-groups of classical type (see §2) we obtain a description of all classical Jordan subgroups. We state this classification result as a theorem.

THEOREM 1.13. *A complete list of the classical Jordan subgroups of connected simple complex algebraic groups with trivial center is given in Table 1 (see page 9). For each such subgroup j of G there are also described in Table 1 the Weyl J-group $W = W(G, j)$ and the J-root system Ω of G relative to j.*

The proof of Theorem 1.13 is given at the end of §4.

A complete description of the nonclassical Jordan subgroups is given in §5.

TABLE 1. Classical Jordan subgroups and their Weyl J-groups

Group G (type)	Jordan subgroup j	J-root system Ω embedded in the vector space J	Weyl J-group W in the representation on J
1	2	3	4
A_{p^n-1} p prime	$\mathbb{Z}_p^{2n}$	$\Omega = J\setminus\{0\}$	$\mathrm{Sp}_{2n}(\mathbb{F}_p)$
B_n $n \geq 3$	$\mathbb{Z}_2^{2n}$	$\Omega = \{e_i + e_j \mid i \neq j,$ $i, j = 1, \ldots, 2n+1\},$ $e_i \in J,\ e_1 + \cdots + e_{2n+1} = 0$	S_{2n+1} (3)
$C_{2^{n-1}}$ $n \geq 2$	$\mathbb{Z}_2^{2n}$	$\Omega = \{\gamma \in J \mid f(\gamma) \neq 0\}$ $f = \sum_{i=1}^n x_{2i-1}x_{2i} + x_1^2 + x_2^2$	$O_f(\mathbb{F}_2)$
D_{n+1} $n \geq 4$	$\mathbb{Z}_2^{2n}$	$\Omega = \{e_i + e_j \mid i \neq j,\ i, j = 1, \ldots, 2n+2\}$ $J = \{\sum x_i e_i \mid \sum x_i = 0\}$ $/\{e_1 + \cdots + e_{2n+2}\}$	S_{2n+2} (4)
$D_{2^{n-1}}$ $n \geq 3$	$\mathbb{Z}_2^{2n}$	$\Omega = \{\gamma \in J / g(\gamma) \neq 0\}$ $g = \sum_{i=1}^{n} x_{q_i-1}x_{2i}$	$O_g(\mathbb{F}_2)$

(3) The representation $S_{2n+1} : J$ is defined by permutations of the vectors $e_1, e_2, \ldots, e_{2n+1}$.

(4) Suppose $S_{2n+2} : \overline{J} = F_2[\overline{e}_1, \ldots, \overline{e}_{2n+2}]$ is a representation of S_{2n+2} by permutations of the basis $\{\overline{e}_i\}$ and $\pi : \overline{J} \to \widetilde{J} = \overline{J}\{\overline{e}_1 + \cdots + \overline{e}_{2n+2}\}$. Put $e_i = \pi(\overline{e}_i)$. Then $J \subset \widetilde{J}$ and J is generated by the vectors $\{e_i + e_j\}$.

§2. Linear groups generated by transvections

The theory of linear groups generated by transvections was developed in connection with the classification of such groups. This classification, under the assumption that the group has no unipotent normal subgroups, was done almost completely by McLaughlin [8], [9], Pollatsek [10], and Zalesskiĭ and Serezhkin [2]. For the present paper we need some of these results, obtained mainly in [8] and [9].

We will state the main definitions and results from the theory of groups generated by transvections needed to study Jordan subgroups. Proofs of most of these results can be found in [8] and [9]. Our terminology is based on our desire to emphasize the parallels between groups generated by transvections and groups generated by reflections. Thus, throughout this section we consider only completely reducible linear groups over a finite prime field $\mathbb{F}_p$ that are generated by transvections. It is known that each transvection r of a finite-dimensional vector space J is defined by a nonzero linear form $\alpha \in J^*$ and a nonzero vector $h \in \operatorname{Ker}\alpha$. Namely, we have

LEMMA 2.1. *Suppose* $\alpha \in J^*$ *and* $h \in J$, *where* $h \neq 0$, $\alpha(h) = 0$. *We define a linear transformation* $r_{\alpha,h}\colon J \to J$ *by the formula*

$$r_{\alpha,h}(x) = x - \alpha(x)h, \qquad x \in J.$$

Then

(1) $r_{\alpha,h}$ *is a transvection*;

(2) *any transvection* r *agrees with one of the transformations* $r_{\alpha,h}$, *i.e.,* $r = r_{\alpha,h}$, $\alpha \in J^*$, $h \in \operatorname{Ker}\alpha$;

(3) *the transformation* $r^*_{\alpha,h}\colon J^* \to J^*$ *conjugate to* $r_{\alpha,h}$ *is defined by the formula*

$$r^*_{\alpha,h} = \omega + \omega(h)\alpha, \qquad \omega \in J^*.$$

The proof of the lemma amounts to a formal verification.

Now consider any linear group W generated by transvections. As in the theory of groups generated by reflections, an important role in the study of W is played by an analogue of a root system:

DEFINITION 2.2. A linear form $\alpha \in J^*$ will be called a T-root relative to W if for some $h \in \operatorname{Ker}\alpha$ the transvection $r_{\alpha,h}$ belongs to W: $r_{\alpha,h} \in W$. The set Ω of all T-roots relative to W will be called the T-root system of W.

We introduce one more standard notation. Suppose α is a T-root relative to W. We put

$$J_{(\alpha)} = \{h \in J \mid r_{\alpha,h} \in W\} \cup \{0\}.$$

LEMMA 2.3. *The set of vectors* $J_{(\alpha)}$ *is a linear subspace of vector space* J.

The proof of the lemma follows from the obvious formulas

$$r_{\alpha,h} \cdot r_{\alpha,g} = r_{\alpha,h+g} \quad \text{and} \quad (r_{\alpha,h})^n = r_{\alpha,nh},$$

where n is any integer.

The subspace $J_{(\alpha)}$ will be called the subspace of roots dual to the J-root α, and the union $\Omega^* = (\bigcup_{\alpha\in\Omega} J_{(\alpha)})\backslash\{0\}$ will be called the dual T-root system.

The study of irreducible groups generated by transvections splits naturally into two cases, depending on whether the group preserves a nonzero bilinear form. It turns out that there is only one series of groups that do not preserve any bilinear form:

PROPOSITION 2.4 (McLaughlin [8], [9]). *Suppose W is an irreducible linear group acting in a vector space J over the field $\mathbb{F}_p$ that is generated by transvections and does not preserve any nonzero bilinear form on J. Then W is the group of all transformations with determinant 1, i.e., $W = \mathrm{SL}(J)$.*

For the group $W = \mathrm{SL}(J)$ it is easy to find the T-root system $\Omega = \Omega(W)$ and the spaces $J_{(\alpha)}$, $\alpha \in \Omega$.

LEMMA 2.5. *Suppose $W = \mathrm{SL}(J)$. Then each nonzero linear form $\alpha \in J^*$ is a T-root for W, i.e., $\Omega = J^*\backslash\{0\}$. For each $\alpha \in \Omega$ the subspace of dual roots $J_{(\alpha)}$ is equal to the kernel of α: $J_{(\alpha)} = \mathrm{Ker}\,\alpha$.*

Now consider the case where W is irreducible and preserves a nonzero, hence nondegenerate, bilinear form $\langle x, y\rangle$.

PROPOSITION 2.6 (McLaughlin [9]). *Suppose $\langle x, y\rangle$ is a nondegenerate bilinear form preserved by W. Then it is skew-symmetric: $\langle x, y\rangle = -\langle y, x\rangle$.*

An important result pertaining to irreducible groups generated by transvections and preserving a nondegenerate bilinear form is, in our terminology, that each such group is completely determined by its T-root system. Before making this assertion precise, we mention the main properties of the T-root system of W.

PROPOSITION 2.7. *Suppose W is an irreducible group generated by transvections and preserving a nondegenerate bilinear form $\langle x, y\rangle$ on the vector space J, and Ω is the T-root system of W. Then*

(1) *the linear span of the set Ω is J^*;*

(2) *if $\alpha, \beta \in \Omega$ and $\langle\alpha, \beta\rangle \neq 0$, then $\alpha + \beta \in \Omega$;*

(3) *if $\alpha \in \Omega$, then $\lambda\alpha \in \Omega$ for any $\lambda \in \mathbb{F}_p^*$;*

(4) *the T-root system Ω is indecomposable* ([5]);

(5) *for each $\alpha \in \Omega$ the subspace $J_{(\alpha)}$ of roots dual to α is one-dimensional: $J_{(\alpha)} = \{\lambda h_\alpha \mid \lambda \in \mathbb{F}_p\}$, where $h_\alpha \in J$ is the vector dual to α: $\langle h_\alpha, x\rangle \equiv \alpha(x)$.*

The proof of assertions (1)–(5) can be found in [8] and [9].

It turns out that properties (1)–(4) completely characterize the T-root system Ω of W. Namely, we have

([5]) A system of vectors $\Omega \subset J^*$ is called indecomposable if there is no decomposition $J^* = J_1 \dotplus J_2$ of the vector space J^* into a direct sum of proper subspaces such that $\Omega = (\Omega\cap J_1)\cup(\Omega\cap J_2)$.

PROPOSITION 2.8. *Suppose J is a vector space over the field $\mathbb{F}_p$, $\langle x, y\rangle$ is a nondegenerate skew-symmetric bilinear form on J, and Ω is a system of vectors in J^* satisfying conditions* (1)–(4) *of Proposition* 2.7. *Then there exists a unique linear group W generated by transvections and preserving the form $\langle x, y\rangle$ such that the T-root system of W is Ω. The group W is an irreducible linear group.*

PROOF. We construct a set of transvections Δ:

$$\Delta = \{r_{\alpha, \lambda\alpha^*} | \alpha \in \Omega, \ \lambda \in \mathbb{F}_p^*\}.$$

Let W be the linear group generated by the set of transformations Δ. Then it is easy to see that W is the desired group. It follows from Proposition 2.7 that it is unique. The proposition is proved.

We will now study the way in which the identical representation of an arbitrary completely reducible group generated by transvections decomposes into a sum of irreducible groups.

PROPOSITION 2.9 (McLaughlin [8], [9]). *Suppose W is a reducible, completely reducible linear group acting in the vector space J and generated by transvections, and Ω is the T-root system of W. Then there exists a decomposition $J = J_0 \dotplus J_1 \dotplus \cdots \dotplus J_m$ of J into a direct sum of linear subspaces J_i with the dual decomposition $J^* = J_0^* \dotplus J_1^* \dotplus \cdots \dotplus J_m^*$, where $J_i^* = \{\omega \in J^* | \omega(J_k) = 0$ when $k \neq i\}$, such that*

(1) $\Omega \cap J_0^* = \varnothing$;

(2) $\Omega = \bigcup_{i=1}^m (\Omega \cap J_i^*)$;

(3) *if $\alpha \in \Omega \cap T_i^*$, then $J_{(\alpha)} \subset J_i$;*

(4) *$W = W_1 \times \cdots \times W_m$, where each of the subgroups W_i acts trivially on the subspaces J_k with $k \neq i$ and is an irreducible subgroup generated by transvections on J_i with T-root system $\Omega \cap J_i^*$.*

We will now state McLaughlin's classification theorem for irreducible groups generated by transvections in the case under consideration, i.e., over the field $\mathbb{F}_p$.

THEOREM 2.10 (McLaughlin [8], [9]). *Suppose W is an irreducible linear group acting in a vector space J over the (finite prime) field $\mathbb{F}_p$ and generated by transvections. If W does not preserve any nonzero bilinear form on J, then $W = \mathrm{SL}(J)$. A complete list of the other groups W is given in column* 4 *of Table* 1. *For each group W the table also indicates its T-root system Ω (column* 3).

It can be seen from Table 1 that if $p \neq 2$, then there are only two series of irreducible linear groups over $\mathbb{F}_p$ generated by transvections: $W = \mathrm{SL}_n(\mathbb{F}_p)$, the special linear group, and $W = \mathrm{Sp}_{2n}(\mathbb{F}_p)$, the symplectic group.

Each group W generated by transvections acts naturally on the set Ω of its T-roots and the dual T-root system $\Omega^* = (\bigcup_{\alpha\in\Omega} J_{(\alpha)})\backslash\{0\}$.

We will now establish some properties of this action.

PROPOSITION 2.11. *Suppose W is an irreducible linear group over $\mathbb{F}_p$ generated by transvections, Ω is the T-root system of W, and Ω^* is the dual T-root system. Then*

(1) *the natural action of W on Ω is transitive;*

(2) *the natural action of W on Ω^* is transitive;*

(3) *if W preserves a nondegenerate bilinear form $\langle x, y\rangle$, then W acts transitively on the subset*

$$\Psi \subset \Omega \times \Omega, \quad \Psi = \{(\alpha, \beta) | \alpha, \beta \in \Omega, \ \langle \alpha, \beta \rangle = 1\}.$$

PROOF. Assertions (1) and (2) were proved by McLaughlin in [9].

Let us prove (3). Since W acts transitively on the set Ω of its T-roots, it suffices to prove that for two pairs of T-roots (α, β) and (α, γ) such that $\langle \alpha, \beta\rangle = \langle \alpha, \gamma\rangle = 1$ there exists an element $w \in W$ such that $w(\alpha) = \alpha$, $w(\beta) = \gamma$. We will construct such an element w. If the vectors β and γ are not orthogonal, i.e., $\langle \beta, \gamma\rangle = \lambda$, where $\lambda \neq 0$, then we put $w = r_{\lambda^{-1}(\beta-\gamma), \lambda^{-1}(\beta-\gamma)^*}$. The transformation w belongs to W, since, by Proposition 2.7, $(\beta - \gamma)$ is a T-root. We can verify that w is the desired element. If the vectors β and γ are orthogonal, i.e., $\langle \beta, \gamma \rangle = 0$, then the element r_{α,α^*} of W sends the pair (α, β) into the pair $(\alpha, \alpha+\beta)$: $r_{\alpha,\alpha^*}(\alpha, \beta) = (\alpha, \alpha+\beta)$. But since $(\alpha+\beta, \gamma) \neq 0$, it follows from what was proved above that there exists an element w^{-1} such that $w^{-1}(\alpha, \alpha+\beta) = (\alpha, \gamma)$. Thus the element $w = w^{-1} r_{\alpha,\alpha^*}$ of W is the desired one. The proposition is proved.

We will now show that W coincides with the group of all symplectic automorphisms of its T-root system.

PROPOSITION 2.12. *Suppose W is an irreducible linear group acting in a vector space J over the field $\mathbb{F}_p$ that is generated by transvections and preserves a nondegenerate bilinear form $\langle x, y\rangle$, and Ω is the T-root system of W. Then the group $\overline{W}$ of all linear transformations of J preserving the form $\langle x, y\rangle$ and sending the T-root system Ω into itself coincides with W: $\overline{W} = W$.*

PROOF. We will prove the proposition by using the classification of the groups W and T-root systems Ω (see Theorem 2.10).

In the case where W is the symplectic group, $W = \mathrm{Sp}_{2n}(\mathbb{F}_p)$, the proposition is obvious.

In the case where W is one of the orthogonal groups over the field $\mathbb{F}_2$, $W = \mathrm{O}_n(\mathbb{F}_2)$, the proposition follows trivially from the description of the T-root system Ω: the set of vectors not in Ω coincides with the zeros of the quadratic form.

Now consider the remaining case where W is the symmetric group, $W = S_n$. It follows from Table 1 that the T-root system Ω is the set of vectors $\Omega = \{e_i + e_j | i, j = 1, \dots, n\}$, and the group $W = S_n$ acts on Ω by permuting the e_i. Suppose $u \in \overline{W}$ is a transformation preserving the bilinear form $\langle x, y\rangle$ and the set Ω of T-roots. By Proposition 2.11, there exists an element $w_0 \in W$ such that $w_0u(e_1+e_2) = e_1+e_2$, $w_0u(e_2+e_3) = e_2+e_3$. Since the transformation w_0u preserves the form $\langle x, y\rangle$, it follows that $w_0u(e_3+e_4) = e_3+e_i$, $i > 3$. But then there exists $w_1 \in W$ such that $w_1w_0u(e_1+e_2) = e_1+e_2$, $w_1w_0u(e_2+e_3) = e_2+e_3$, $w_1w_0u(e_3+e_4) = e_3+e_4$. Repeating this argument, we obtain $w_k \cdots w_0u = e$. Thus $u \in W$. The proposition is proved.

To conclude this section we mention two simple lemmas that will be needed for the subsequent exposition (§3).

LEMMA 2.13. *Suppose W is a completely reducible linear group generated by transvections and acting in a vector space J over the field $\mathbb{F}_p$, and α is any T-root. Let $W_\alpha = \{w \in W | W^*(\alpha) = \lambda\alpha,\ \lambda \in \mathbb{F}_p\}$ and, as usual, $J_{(\alpha)} = \{h \in J | r_{\alpha,h} \in W\} \cup \{0\}$. Then the group W_α acts transitively on the set of vectors $J_{(\alpha)}\backslash\{0\}$.*

PROOF. In view of Proposition 2.9, it suffices to prove the lemma for an irreducible group W. If W is irreducible and preserves a nondegenerate bilinear form, then, by Proposition 2.7, $J_{(\alpha)} = \{\lambda\alpha^* | \lambda \in \mathbb{F}_p\}$. The assertion of the lemma now obviously follows from the transitivity of the group on W (Proposition 2.11). If W does not preserve any nonzero bilinear form, then, by Proposition 2.4, $W = \mathrm{SL}(J)$. In this case the assertion of the lemma is obvious. The lemma is proved.

LEMMA 2.14. *Suppose W is a completely reducible linear group generated by transvections and acting in a vector space J over the field $\mathbb{F}_p$, and α, β are two T-roots of W with subspaces $J_{(\alpha)}$, $J_{(\beta)}$ of dual roots. If $\alpha(J_{(\beta)}) = \beta(J_{(\alpha)}) = 0$ and $J_{(\alpha)} \cap J_{(\beta)} \neq \{0\}$, then the T-roots α, β are proportional: $\alpha = \lambda\beta$, $\lambda \in \mathbb{F}_p^*$.*

It is clear that it suffices to prove the lemma for an irreducible group W and then apply Proposition 2.9. Suppose the group is irreducible and preserves a nonzero bilinear form. Then, by Proposition 2.7, $J_{(\alpha)} = \{\lambda\alpha^* | \lambda \in \mathbb{F}_p\}$, and the assertion of the lemma is obvious. If W is irreducible and does not preserve any nonzero bilinear form, then $W = \mathrm{SL}(J)$ (Proposition 2.4) and the assertion of the lemma can be verified trivially. The lemma is proved.

§3. Structure of a simple algebraic group associated with a Jordan subgroup

The main purpose of this section is to prove Theorem 1.4, which was stated in §1 and says that the Weyl J-group is generated by transvections. The proof follows from a detailed study of the structure of G, more precisely, of the root subgroups $G_{(\alpha)}$ relative to a Jordan subgroup $j \subset G$, the J-root system Ω, and the root decomposition $\mathfrak{J} = \dot{\Sigma}_{\alpha\in\Omega}\mathfrak{J}_\alpha$ of the Lie algebra $\mathfrak{J}$ of G. Along the way, we will explicitly construct a bilinear form $\langle x, y\rangle$ preserved by the Weyl J-group W and explain when it is nondegenerate.

All of the main steps leading to the proof of Theorem 1.4 will be stated as propositions.

We first introduce the definition of a root subgroup.

DEFINITION 3.1. Suppose j is a Jordan subgroup of G, α is a J-root, and $k_{(\alpha)}$ is the subgroup of j corresponding to the kernel of α in J under the standard isomorphism $\varepsilon\colon J \simeq j$, i.e., $k_{(\alpha)} = \varepsilon(\operatorname{Ker}\alpha)$. Then the connected centralizer $G_{(\alpha)} = Z_G(k_{(\alpha)})^0$ will be called the root subgroup corresponding to the J-root α.

The root subgroup $G_{(\alpha)}$ defined analogously in the Cartan theory is an almost direct product of a three-dimensional simple subgroup SL_2 and a torus contained in the Cartan subgroup. In the theory of Jordan subgroups, a root subgroup is always a torus:

PROPOSITION 3.2. *Suppose j is a Jordan subgroup of G, α is a J-root, and $G_{(\alpha)}$ is the corresponding root subgroup. Then $G_{(\alpha)}$ is a torus.*

PROOF. The Jordan subgroup j obviously acts on $G_{(\alpha)}$ by conjugation, since $aG_{(\alpha)}a^{-1} = G_{(\alpha)}$ for each $a \in j$. Moreover, this action can be reduced to that of a cyclic group of prime order p, namely, the group $j/k_{(\alpha)}$. The set $z_{(\alpha)}$ of elements of $G_{(\alpha)}$ invariant under the action of $j/k_{(\alpha)}$ on $G_{(\alpha)}$ obviously coincides with the intersection of the centralizer $Z_G(j)$ with $G_{(\alpha)}$: $z_{(\alpha)} = Z_G(j) \cap G_{(\alpha)}$. Since $|Z_G(j)| < \infty$, it follows that $|z_{(\alpha)}| < \infty$. It is known that if a connected semisimple complex algebraic group admits an automorphism of prime order whose set of fixed points is finite, then it is a torus (see, e.g., [3]). Thus $G_{(\alpha)}$ is a torus. The proposition is proved.

The first step in proving Theorem 1.4 is to construct a sufficiently large number of transvections in the Weyl J-group $W = W(G, j)$. We will look for these transvections (more precisely, for their preimages in the normalizer $N = N_G(j)$ of the Jordan subgroup j) in the intersections $N_{(\alpha)} = N \cap G_{(\alpha)}$ of the normalizer N of the Jordan subgroup with the root subgroups $G_{(\alpha)}$.

PROPOSITION 3.3. *Suppose j is a Jordan subgroup of G, α is any J-root, and $j_{(\alpha)}$ is the intersection of the Jordan subgroup j with the root subgroup $G_{(\alpha)}$: $j_{(\alpha)} = j \cap G_{(\alpha)}$. Then for each $a \in j_{(\alpha)} \setminus \{e\}$ there exists an element $r \in N_{(\alpha)} = N_G(j) \cap G_{(\alpha)}$ sent into the transvection $r_{\alpha,\varepsilon^{-1}(a)}$ belonging to the Weyl J-group W: $\overline{r} = r_{\alpha,\varepsilon^{-1}(a)}$ (as before, ε denotes the standard isomorphism $\varepsilon\colon J \simeq j$ of the additive group of the vector space J associated with j and the group j itself).*

The proof of the proposition is based on the following important lemma on an automorphism of a torus.

LEMMA 3.4. *Suppose a is an automorphism of prime order p of a complex algebraic torus T such that its set S of fixed points is finite. Then*

(1) *the torus* T *is a direct product* $T = T_1 \times \cdots \times T_k$ *of tori* T_i, *each of which is a-invariant and has dimension* $p-1$: $\dim T_i = p-1$;

(2) $S = S_1 \times \cdots \times S_k$, *where* $S_i = S \cap T_i$, *and* $S_i \simeq \mathbb{Z}_p$ *for each* $i = 1, \dots, k$;

(3) *the Lie algebra* $\mathscr{L}_i$ *of the torus* T_i *is a direct sum* $\mathscr{L}_i = \sum_{j=1}^{p-1} \mathscr{L}_{ij}$ *of one-dimensional subspaces* $\mathscr{L}_{ij}$ *such that* $a(X) = \varepsilon^j X$ *for each* $X \in \mathscr{L}_{ij}$ (ε *is a fixed primitive* pth *root of unity*).

PROOF. In view of the duality of the torus T and its character lattice $\mathfrak{X}$, the study of the automorphism a is equivalent to the study of the conjugate automorphism $a^*: \mathfrak{X} \to \mathfrak{X}$. This, in turn, is equivalent to the study of the integral representation of the cyclic group $\mathbb{Z}_p \cong \{e, a^*, \dots, a^{*p-1}\}$ of prime order p in the free $\mathbb{Z}$-module $\mathfrak{X}$. Thus to prove the lemma it suffices to reformulate it in terms of the representation $\mathbb{Z}_p: \mathfrak{X}$, and in this form the lemma becomes an easy consequence of known results in the theory of integral $\mathbb{Z}_p$-modules (see, e.g., [4]). We omit the details. The lemma is proved.

We will now apply Lemma 3.4 to an automorphism of the root subgroup $G_{(\alpha)}$. Namely, suppose α is any J-root and $k_{(\alpha)}$ is the subgroup of the Jordan subgroup j corresponding to the kernel of the linear form α, $k_{(\alpha)} = \varepsilon(\operatorname{Ker}\alpha)$, $G_{(\alpha)}$ is the root subgroup, and, finally, a_α is an element of j not in $k_{(\alpha)}$, $a_\alpha \in j \backslash k_{(\alpha)}$. By what has been proved, $G_{(\alpha)}$ is a torus. The element a_α acts on $G_{(\alpha)}$ by the rule $g \mapsto a_\alpha g a_\alpha^{-1}$. The order of a_α is the prime p. The set $z_{(\alpha)}$ of fixed points in the torus $G_{(\alpha)}$ under the automorphism a_α coincides with the intersection $Z_G(j) \cap G_{(\alpha)}$ of the centralizer $Z_G(j)$ of the Jordan subgroup with $G_{(\alpha)}$: $z_{(\alpha)} = Z_G(j) \cap G_{(\alpha)}$. Consequently, $|z_{(\alpha)}| < \infty$ and all conditions of Lemma 3.4 for the automorphism a_α of the torus $G_{(\alpha)}$ are satisfied.

We now obtain two direct corollaries of the lemma.

COROLLARY 3.5. *Suppose* j *is a Jordan subgroup of* G *and* $\alpha \in \Omega$ *is any* J*-root. Then there exists a natural number* m *such that*

(1) *the dimension of the root subgroup* $G_{(\alpha)}$ *is* $m(p-1)$: $\dim_{\mathbb{C}} G_{(\alpha)} = m(p-1)$;

(2) *the subgroup* $z_{(\alpha)} = Z_G(j) \cap G_{(\alpha)}$ *is isomorphic to the direct product of* m *copies of* $\mathbb{Z}_p$: $z_{(\alpha)} \simeq \mathbb{Z}_p^m$;

(3) *for each* $\lambda \in \mathbb{F}_p^*$, $\lambda\alpha$ *is also a* J*-root and* $\dim_{\mathbb{C}} \mathfrak{J}_{\lambda\alpha} = \dim_{\mathbb{C}} \mathfrak{J}_\alpha$.

In proving Proposition 3.3 we will use another corollary of Lemma 3.4.

COROLLARY 3.6. *Suppose* j *is a Jordan subgroup of the Lie group* G, α *is any* J*-root, and* $a_\alpha \in j \backslash k_{(\alpha)}$, *where* $k_{(\alpha)} = \varepsilon(\operatorname{Ker}\alpha)$. *Then for each element* $b \in z_{(\alpha)}$ *of the group* $z_{(\alpha)} = Z_G(j) \cap G_{(\alpha)}$ *there exists an element* $r \in G_{(\alpha)}$ *of the root subgroup* $G_{(\alpha)}$ *such that* $r a_\alpha r^{-1} = a_\alpha b$.

PROOF. Since the element b of the torus $G_{(\alpha)}$ is invariant under the automorphism $a_\alpha: g \mapsto a_\alpha g a_\alpha^{-1}$ of $G_{(\alpha)}$, it follows that a_α induces an automorphism $\overline{a}_\alpha$ of the factor group $G_{(\alpha)}/\{b\}$, $\overline{a}_\alpha: G_{(\alpha)}/\{b\} \to G_{(\alpha)}/\{b\}$. By Lemma 3.4, the number of fixed points in the torus $G_{(\alpha)}/\{b\}$ under the automorphism $\overline{a}_\alpha$ is equal to the number of fixed points in the torus $G_{(\alpha)}$ under the automorphism a_α. Since the set $z_{(\alpha)}$ of fixed points under the automorphism $a_\alpha: G_{(\alpha)} \to G_{(\alpha)}$ is projected into the torus $G_{(\alpha)}/\{b\}$ nonisomorphically, $z_{(\alpha)} \to z_{(\alpha)}/\{b\}$, it follows that the torus $G_{(\alpha)}$ contains an element r_0 that does not lie in $z_{(\alpha)}$, $r_0 \notin z_{(\alpha)}$, but whose image under the projection $G_{(\alpha)} \to G_{(\alpha)}/\{b\}$ is fixed under $\overline{a}_\alpha$. This means that $a_\alpha r_0 a_\alpha^{-1} = r_0 b^k$, $(k, p) = 1$. It is clear that a suitable power r_0^l of r_0 will serve as the desired element r, $r = r_0^l$. The corollary is proved.

PROOF OF PROPOSITION 3.3. Suppose a is any nonidentity element of the subgroup $j_{(\alpha)} = j \cap G_{(\alpha)}$, $a \in j_{(\alpha)}$. Since we obviously have $j_{(\alpha)} \subset z_{(\alpha)}$, it follows from Corollary 3.6 that there exists an element $r \in G_{(\alpha)}$ in the root subgroup $G_{(\alpha)}$ such that $r a_\alpha r^{-1} = a_\alpha a$. Since r commutes with the subgroup $k_{(\alpha)} = \varepsilon(\operatorname{Ker}\alpha)$ of the Jordan subgroup j, and j is generated by $k_{(\alpha)}$ and the element a_α, it follows that $r \in N_G(j)$. Let $\overline{r} = A(r)$ be the image of r under the projection $A: N_G(j) \to W$. It is clear that $\overline{r}$, as a linear transformation of the vector space J associated with j, agrees with the transvection $r_{\alpha, \varepsilon^{-1}(a)}$, i.e., $\overline{r} = r_{\alpha, \varepsilon^{-1}(a)}$. The proposition is proved.

The next step in constructing transvections in the Weyl J-group W is to prove the groups $j_{(\alpha)}$ are nontrivial.

PROPOSITION 3.7. *Suppose j is a Jordan subgroup of the Lie group G and α is any J-root. Then the intersection $j_{(\alpha)} = j \cap G_{(\alpha)}$ of j with the root subgroup $G_{(\alpha)}$ is nontrivial*: $j_{(\alpha)} \neq \{e\}$.

To prove Proposition 3.7 we require several lemmas.

LEMMA 3.8. *Suppose j is a Jordan subgroup of the Lie group G and z is the centralizer of j in G, $z = Z_G(j)$. Then the normalizers of z and j in G coincide*: $N_G(z) = N_G(j)$.

PROOF OF THE LEMMA. Consider the normal subgroup k of $N_G(z)$ generated by the subgroup j: $k = \{gjg^{-1} | g \in N_G(z)\}$. It is clear that k is contained in the center of $z = Z_G(j)$. Therefore k is a diagonalizable subgroup of G. In view of the condition of the maximality of $N_G(j)$ imposed in the definition of a Jordan subgroup, $N_G(k) = N_G(j)$. Since k is normal in $N_G(z)$ by construction, it follows that $N_G(z) = N_G(j)$. The lemma is proved.

LEMMA 3.9. *Suppose j is a Jordan subgroup of the Lie group G, α is any J-root, $G_{(\alpha)}$ is the root subgroup, and $j_{(\alpha)} = j \cap G_{(\alpha)}$, $z_{(\alpha)} = Z_G(j) \cap G_{(\alpha)}$.*

Then $j_{(\alpha)}$ *is the intersection of the center* c *of the group* $z = Z_G(j)$ *with* $z_{(\alpha)}$, *i.e.,* $j_{(\alpha)} = \{x \in z_{(\alpha)} | xgx^{-1} = g \ \forall g \in z\}$.

PROOF OF THE LEMMA. Suppose $x \in z_{(\alpha)}$ and x commutes with all elements of z. By Corollary 3.6, there exists an element r such that $ra_\alpha r^{-1} = a_\alpha x$ for some $a_\alpha \in j$ that does not lie in the kernel $k_{(\alpha)}$ of the root α, $a_\alpha \in j \backslash k_{(\alpha)}$. It is clear that the subgroup rjr^{-1}, like j, is contained in the center c of z. Thus $r \in N_G(j)$. By Lemma 3.8, $N_G(j) = N_G(z)$, hence $r \in N_G(j)$. But then x lies in j, inasmuch as $x = a_\alpha^{-1} r a_\alpha r^{-1}$. The lemma is proved.

LEMMA 3.10. *Suppose* k *is a group of automorphisms of a connected reductive complex algebraic group* G *and is isomorphic to a direct product of cyclic groups of prime order* p, $k \simeq \mathbb{Z}_p \times \cdots \times \mathbb{Z}_p$. *Then the factor group* $Z_G(k)/Z_G(k)^0$ *is a p-group (here* $Z_G(k)$ *is the set of elements of* G *fixed under all automorphisms in* k).

PROOF. The proof of Lemma 3.10, by induction on the p-rank of the group k, reduces easily to the case where k is cyclic, $k \simeq \mathbb{Z}_p$. For a cyclic group the assertion of the lemma was proved in [7]. The lemma is proved.

PROOF OF PROPOSITION 3.7. The subgroup $z_{(\alpha)} = Z_G(j) \cap G_{(\alpha)}$ is obviously a normal subgroup of $z = Z_G(j)$. By Lemma 3.10, z is a p-group, hence by the usual argument from the theory of p-groups, applied to the action of z on $z_{(\alpha)}$ by conjugation, we can prove that $z_{(\alpha)}$ contains a nonidentity element $x \in z_{(\alpha)}$ that commutes with all elements of z. By Lemma 3.9, $x \in j_{(\alpha)}$. Consequently, $j_{(\alpha)} \neq \{e\}$. The proposition is proved.

Thus we have constructed many transvections in the Weyl J-group W. Now consider the subgroup W_t of W generated by all transvections lying in W. It is clear that W_t is a normal subgroup of W. It follows from the irreducibility condition that W is an irreducible linear group acting in the vector space J associated with j. Thus the normal subgroup W_t is a completely reducible linear group.

We will now study the connections between certain invariants of the group W_t and the Jordan subgroup j of G.

PROPOSITION 3.11. *Suppose* j *is a Jordan subgroup of* G *and* W_t *is the subgroup of the Weyl* J*-group* W *generated by all transvections lying in* W. *Then*

(1) *the* J*-root system* Ω *of* G *relative to* j *is the same as the* T*-root system* Ω_t *relative to* W_t;

(2) *for each* $\alpha \in \Omega$ *the subspace* $J_{(\alpha)} = \{x \in J | r_{\alpha, x} \in W_t\} \cup \{0\}$ *of* J *introduced in* §2 *is the image of the subgroup* $j_{(\alpha)} = j \cap G_{(\alpha)}$ *under the isomorphism* $\varepsilon^{-1}: j \simeq J$, $J_{(\alpha)} = \varepsilon^{-1}(j_{(\alpha)})$.

PROOF. Recall that a linear form α on the vector space J is called a T-root relative to W_t if there exists a vector $h \in \operatorname{Ker} \alpha$ such that the transvection $r_{\alpha,h}$ lies in W_t, $r_{\alpha,h} \in W_t$ (Definition 2.2). It follows from Propositions 3.3 and 3.7 that for each J-root α there exists $h \in \varepsilon^{-1}(j)$ such that $r_{\alpha,h} \in W_t$. Thus $\alpha \in \Omega_t$, where Ω_t is the T-root system relative to W_t, and $\Omega \subset \Omega_t$. The set of J-roots Ω is obviously invariant under the natural action of W_t in the conjugate space J^*. The group W_t is a completely reducible linear group. Thus the T-root system Ω_t decomposes into several components, on each of which W_t acts transitively (Propositions 2.9 and 2.11). But since the set Ω of all J-roots generates the whole linear space J^*, it follows that $\Omega = \Omega_t$. Our assertion is proved.

By Proposition 3.3, we have the inclusion of subspaces $\varepsilon^{-1}(j_{(\alpha)}) \subset J_{(\alpha)}$, where $j_{(\alpha)} = j \cap G_{(\alpha)}$, $J_{(\alpha)} = \{h \in J | r_{\alpha,h} \in W\}$. By Proposition 3.7, $j_{(\alpha)} \neq \{e\}$. Let $W_{(\alpha)} = \{w \in W_t | w^* \alpha = \lambda \alpha\}$. The subspace $\varepsilon^{-1}(j_{(\alpha)})$ is obviously invariant under $W_{(\alpha)}$. But by Lemma 2.13, the group $W_{(\alpha)}$ is transitive on $J_{(\alpha)} \setminus \{0\}$. Thus $\varepsilon^{-1}(j_{(\alpha)}) = J_{(\alpha)}$. The proposition is proved.

The next step is studying the Weyl J-group is the following proposition.

PROPOSITION 3.12. *The subgroup W_t is an irreducible linear group.*

PROOF. Assume the linear group W_t is reducible. Then $J = J_1 \dot{+} J_2$, where J_1 and J_2 are nonzero W_t-invariant subspaces. By Proposition 2.9, the T-root system Ω_t decomposes: $\Omega_t = (\Omega_t \cap J_1) \cup (\Omega_t \cap J_2)$. Since Ω_t is equal to the J-root system Ω, $\Omega_t = \Omega$, we can construct a decomposition $\mathfrak{J} = \mathfrak{J}_1 \dot{+} \mathfrak{J}_2$ of the complex Lie algebra $\mathfrak{J}$ of G into a direct sum of subspaces $\mathfrak{J}_1$ and $\mathfrak{J}_2$ by putting $\mathfrak{J}_i = \dot{\Sigma}_{\alpha \in \Omega \cap J_i} \mathfrak{J}_\alpha$, $i = 1, 2$.

We will prove that $[\mathfrak{J}_1, \mathfrak{J}_2] = \{0\}$. To do this we must study the arrangement of the root subspaces.

PROPOSITION 3.13. *Suppose j is a Jordan subgroup of the Lie group G, $\mathfrak{J} = \dot{\Sigma}_{\alpha \in \Omega} \mathfrak{J}_\alpha$ is the grading of the Lie algebra $\mathfrak{J}$ associated with j, and α, β are any two J-roots. Then the following two assertions are equivalent*:

(i) $[\mathfrak{J}_\alpha, \mathfrak{J}_\beta] = 0$;

(ii) $\alpha|_{J_{(\beta)}} \equiv \beta|_{J_{(\alpha)}} \equiv 0$.

We will first complete the proof of Proposition 3.12, relying on Proposition 3.13. It follows from 3.13 that $[\mathfrak{J}_1, \mathfrak{J}_2] = \{0\}$. But obviously $[\mathfrak{J}_i, \mathfrak{J}_i] \subset \mathfrak{J}_i$, $i = 1, 2$. Thus $\mathfrak{J}_1$ and $\mathfrak{J}_2$ are ideals of the Lie algebra, which contradicts the fact that the Lie algebra $\mathfrak{J}$ is simple. Therefore W_t is an irreducible linear group. The proposition is proved.

The proof of Proposition 3.13 will be preceded by two lemmas.

LEMMA 3.14. *Suppose j is a Jordan subgroup of the Lie group G, α is any J-root, $\mathfrak{J}_\alpha$ is the weight subspace in the complex Lie algebra of G relative to j corresponding to the J-root α, and $\mathfrak{J}_{(\alpha)}$ is the Lie subalgebra of the* *root subgroup* $G_{(\alpha)}$ *of* G. *Then* $[X, \mathfrak{J}_\alpha] = 0$ *for a vector* $X \in \mathfrak{J}$ *implies* $[X, \mathfrak{J}_{(\alpha)}] = 0$.

Proof. It follows from Lemma 3.4, applied to the torus $G_{(\alpha)}$, that $\mathfrak{J}_{(\alpha)}$ is the smallest algebraic subalgebra containing $\mathfrak{J}_\alpha$. But for such a subalgebra the assertion of the lemma is obvious. The lemma is proved.

Lemma 3.15. *Suppose* j *is a Jordan subgroup of the Lie group* G, α *and* β *are two* J*-roots,* $\mathfrak{J}_\alpha$ *and* $\mathfrak{J}_\beta$ *are the corresponding weight subspaces in the Lie algebra* $\mathfrak{J}$ *relative to the representation of* j *by adjoint operators, and* $X \in \mathfrak{J}_\alpha$. *Then* $[X, \mathfrak{J}_\beta] = 0$ *implies* $[X, \mathfrak{J}_{\lambda\alpha+\mu\beta}] = 0$ *for any* $\lambda, \mu \in \mathbb{F}_p$.

Proof. Since $[\mathfrak{J}_\alpha, \mathfrak{J}_\gamma] \subset \mathfrak{J}_{\alpha+\gamma}$ for any J-root γ, we have $[X, \mathfrak{J}_{\beta+\lambda\alpha}] \subset \mathfrak{J}_{\beta+(\lambda+1)\alpha}$ for any $\lambda \in \mathbb{F}_p$ and $X \in \mathfrak{J}_\alpha$. It now follows from the condition $\operatorname{ad} X(\mathfrak{J}_\beta) = [X, \mathfrak{J}_\beta] = 0$ that the restriction $\operatorname{ad} X|_V$ of the operator $\operatorname{ad} X$ to the invariant subspace $V = \Sigma^{\cdot}_{\lambda\in F_p} \mathfrak{J}_{\beta+\lambda\alpha}$ is nilpotent. But $\operatorname{ad} X$ is a semisimple operator, inasmuch as the vector X lies in the Lie algebra $\mathfrak{J}_{(\alpha)}$ of the root torus $G_{(\alpha)}$. Thus $\operatorname{ad} X|_V = 0$. Consequently, $[X, \mathfrak{J}_{\beta+\lambda\alpha}] = 0$ for any $\lambda \in \mathbb{F}_p$. If we now apply Lemma 3.14 to the element $X \in \mathfrak{J}_\alpha$ and to each of the subspaces $\mathfrak{J}_{\beta+\lambda\alpha}$, we obtain the desired assertion. The lemma is proved.

Proof of Proposition 3.13. Suppose (i) holds: $[\mathfrak{J}_\alpha, \mathfrak{J}_\beta] = 0$. It then follows from Lemma 3.14 that $[\mathfrak{J}_{(\alpha)}, \mathfrak{J}_{(\beta)}] = 0$. Thus the root subgroups $G_{(\alpha)}$ and $G_{(\beta)}$ commute. Since, by Proposition 3.3, $\varepsilon(J_{(\alpha)}) = j \cap G_{(\alpha)}$, we see that the subgroups $j_{(\alpha)} = \varepsilon(J_{(\alpha)})$ and $G_{(\beta)}$ commute. This means that $\beta|_{J(\alpha)} = 0$. Analogously, $\alpha|_{J_{(\beta)}} = 0$. Thus (ii) holds.

Conversely, suppose (ii): $\alpha|_{J_{(\beta)}} = \beta|_{J_{(\alpha)}} = 0$. Assume $[\mathfrak{J}_\alpha, \mathfrak{J}_\beta] \neq 0$. Using this assumption, we will prove that $j_{(\alpha)} \cap j_{(\beta)} \neq \{e\}$, where $j_{(\alpha)} = j \cap G_{(\alpha)}$, $j_{(\beta)} = j \cap G_{(\beta)}$. Then $J_{(\alpha)} \cap J_{(\beta)} \neq \{0\}$. By Lemma 2.14, $\beta = \lambda\alpha$, $\lambda \in \mathbb{F}_p^*$. But since $\mathfrak{J}_\alpha \subset \mathfrak{J}_{(\alpha)}$, $\mathfrak{J}_{\lambda\alpha} \subset \mathfrak{J}_{(\alpha)}$, and $\mathfrak{J}_{(\alpha)}$ is the Lie algebra of the torus $G_{(\alpha)}$, we have $[\mathfrak{J}_\alpha, \mathfrak{J}_\beta] = 0$. This contradicts our assumption. Thus $[\mathfrak{J}_\alpha, \mathfrak{J}_\beta] = 0$, as required.

It remains to prove that $j_{(\alpha)} \cap j_{(\beta)} \neq \{e\}$ if $[\mathfrak{J}_\alpha, \mathfrak{J}_\beta] \neq 0$. Consider the subalgebra $\mathscr{L} = \Sigma^{\cdot}_{\lambda, \mu\in F_p} \mathfrak{J}_{\lambda\alpha+\mu\beta}$ of the Lie algebra $\mathfrak{J}$ of G. The subalgebra $\mathscr{L}$ obviously corresponds to the subgroup $H = Z_G(k_{(\alpha)} \cap k_{(\beta)})^0$, where $k_{(\alpha)} = \varepsilon(\operatorname{Ker} \alpha)$, $k_{(\beta)} = \varepsilon(\operatorname{Ker} \beta)$. Consequently, the root tori $G_{(\alpha)}$ and $G_{(\beta)}$ lie in H: $G_{(\alpha)}, G_{(\beta)} \subset H$. By assumption, $[\mathfrak{J}_\alpha, \mathfrak{J}_\beta] \neq \{0\}$, i.e., the group H is noncommutative. Thus $H = S \cdot T$, where S is a semisimple group (with Lie algebra $\mathfrak{S}$), T is a torus (with Lie algebra $\mathfrak{T}$), and the product of S and T is almost direct. Since each of S and T is invariant under conjugation by elements of the Jordan subgroup j, we have $G_{(\alpha)} = S_{(\alpha)} \cdot T_{(\alpha)}$, where $S_{(\alpha)} = S \cap G_{(\alpha)}$ and $T_{(\alpha)} = T \cap G_{(\alpha)}$. The same, of course, is true for $G_{(\beta)}$: $G_{(\beta)} = S_{(\beta)} \cdot T_{(\beta)}$. We will now prove that $S_{(\alpha)}$ and $S_{(\beta)}$ are maximal tori in S. Suppose $\mathfrak{z}$ is a Lie subalgebra of the centralizer $Z_S(S_{(\alpha)})$. Since we obviously have $\operatorname{Ad} j(\mathfrak{z}) = \mathfrak{z}$, it follows that $\mathfrak{z} = \dot{\sum}_{\lambda, \mu \in F_p} \mathfrak{z}_{\lambda\alpha+\mu\beta}$, where $\mathfrak{z}_{\lambda\alpha+\mu\beta} = \mathfrak{z} \cap \mathfrak{J}_{\lambda\alpha+\mu\beta}$. Suppose $X \in \mathfrak{z}_{\lambda\alpha+\mu\beta}$ and $\mu \neq 0$. Since, by construction, $[X, \mathfrak{z}] = 0$ and $[X, \mathfrak{F}] = 0$, we see that $[X, \mathfrak{J}_\alpha] = 0$. Therefore, by Lemma 3.15, $[X, \mathfrak{J}_{\lambda\alpha+\mu\beta}] = 0$ for any $\lambda, \mu \in \mathbb{F}_p$. Thus $[X, \mathfrak{Z}] = 0$ and, since $\mathfrak{Z}$ is a semisimple algebra, $X = 0$. It follows from what has been proved that $\mathfrak{z} = \mathfrak{Z}_{(\alpha)}$. Thus $S_{(\alpha)}$ is a maximal torus in S. Analogously, $S_{(\beta)}$ is a maximal torus in S.

We will now prove that $j \cap S_{(\alpha)} \neq \{e\}$ and $j \cap S_{(\beta)} \neq \{e\}$. The tori $S_{(\alpha)}$ and $S_{(\beta)}$ are obviously invariant under conjugation by elements of the subgroup $z = Z_G(j)$. Since z is a p-group (according to Lemma 3.10), the torus $S_{(\alpha)}$ contains a nonidentity element a such that $gag^{-1} = a$ for each $g \in z$. By Lemma 3.9, $a \in j$. Thus $j \cap S_{(\alpha)} \neq \{e\}$. Analogously, $j \cap S_{(\beta)} \neq \{e\}$. But it follows from $\alpha(J_{(\beta)}) \equiv \beta(J_{(\alpha)}) \equiv 0$ that the subgroup $j_{(\alpha)} = j \cap G_{(\alpha)}$ and the subgroup $j_{(\beta)} = j \cap G_{(\beta)}$ lie in the center of $H = S \cdot T$. Therefore the subgroups $j \cap S_{(\alpha)}$ and $j \cap S_{(\beta)}$ lie in the center of S. But each maximal torus of S contains the center of S. Thus $S_{(\alpha)} \supset j \cap S_{(\beta)}$ and $S_{(\beta)} \supset j \cap S_{(\alpha)}$. Therefore $j \cap S_{(\alpha)} = j \cap S_{(\beta)}$. Thus $j_{(\alpha)} \cap j_{(\beta)} \neq \{e\}$, as required to complete the proof of the proposition.

Our last step in developing the structure theory is to prove that the Weyl J-group W coincides with the subgroup W_t generated by transvections.

PROPOSITION 3.16. *The subgroup W_t coincides with the whole group W, $W_t = W$.*

The proof of the proposition will be different for Weyl J-groups of classical and nonclassical types.

We will first prove Proposition 3.16 for classical Jordan subgroups, i.e., where the Weyl J-group W preserves a nonzero bilinear form (Definition 1.10). Suppose $\langle x, y\rangle$ is a nonzero W-invariant bilinear form on J. Since this form is preserved by the irreducible group W_t generated by transvections, it follows from Proposition 2.6 that the form $\langle x, y\rangle$ is nondegenerate and skew-symmetric. Since W obviously preserves the J-root system Ω under the action in J^* of conjugate operators, it follows from Proposition 2.12 that $W \subset W_t$. But since W_t is a subgroup of W, we have $W = W_t$, and Proposition 3.16 is proved for classical Jordan subgroups.

To prove Proposition 3.16 in the nonclassical case we will explicitly construct a certain bilinear form $c(x, y)$ on the linear space J and prove that any bilinear form invariant under all transvections $r_\alpha \in W$ is proportional to $c(x, y)$.

Roughly speaking, the form $c(x, y)$ is defined as the commutator of preimages of elements of the Jordan subgroup j in a simply connected covering group $\widehat{G}$ of G. Bilinearity of the mapping follows automatically from known group identities. The only interesting thing to verify is that the range of the indicated mapping is contained in $\mathbb{Z}_p$.

More precisely, suppose $\pi: \widehat{G} \to G$ is a universal covering and Δ is the center of the simply connected covering group $\widehat{G}$ of G. For any $a, b \in j$ we put $c(a, b) = \hat{a}\hat{b}\hat{a}^{-1}\hat{b}^{-1}$, where $\hat{a} \in \pi^{-1}(a)$, $\hat{b} \in \pi^{-1}(b)$.

LEMMA 3.17. *For any $a, b \in j$ the element $c(a, b)$ does not depend on the choice of preimages $\hat{a} \in \pi^{-1}(a)$, $\hat{b} \in \pi^{-1}(b)$ and lies in the center Δ of $\widehat{G}$. Moreover, the following identities hold*:

(1) $c(a, b) = c(b, a)^{-1}$;
(2) $c(a, bd) = c(a, b)c(a, d)$;
(3) $c(a, b^k) = c(a, b)^k$, $k \in \mathbb{Z}$;
(4) $c(gag^{-1}, gbg^{-1}) = c(a, b)$ *for any* $g \in N_G(j)$.

The proof of the lemma amounts to a formal verification of all assertions.

LEMMA 3.18. *The range f of the mapping $c: j \times j \to \Delta$ either consists of the identity element, $f = \{e\}$, or is a subgroup isomorphic to $\mathbb{Z}_p$, $f = \mathbb{Z}_p$.*

PROOF. By Lemma 3.17, $c(a, b)^p = c(a, b^p) = c(a, e) = e$. Thus the image f of the mapping c is a subgroup of period p of the group Δ. It is known that either the center Δ of a simply connected, simple group $\widehat{G}$ is cyclic, or $\widehat{G} = D_{2n}$ and $\Delta \simeq \mathbb{Z}_2 \times \mathbb{Z}_2$. In the first case it is clear that $f = \{e\}$ or $f = \mathbb{Z}_p$. In the second, according to Lemma 3.17, the mapping $C: J \times J \to \Delta$ obtained from $c: j \times j \to \Delta$ via the isomorphism $\varepsilon^{-1}: j \simeq J$ is a bilinear mapping preserved by W. By what has been proved, the group W is irreducible. Thus the rank of the image f of the mapping c is 1 or 0. Consequently, $f = \mathbb{Z}_2$ or $f = \{e\}$. The lemma is proved.

Using the mapping $c: j \times j \to f \subset \Delta$ constructed above, we will now construct a bilinear form $c(x, y)$ with values in $\mathbb{F}_p$ on the vector space J associated with the Jordan subgroup j. The mapping $c: j \times j \to f$ can be carried over via the isomorphism $\varepsilon^{-1}: j \simeq J$ to a mapping $c: J \times J \to F$, where F is constructed from f as was J from j, i.e., $F = \mathrm{Hom}(f^\#, \mathbb{F}_p)$, where $f^\# = \mathrm{Hom}(f, \mathbb{C}^*)$. It is clear that F is a one-dimensional vector space over $\mathbb{F}_p$ if $f \neq \{e\}$, and $F = \{0\}$ if $f = \{e\}$. We identify F in some way with $\mathbb{F}_p$ in the case where $F \neq \{0\}$, $\tau: F \simeq \mathbb{F}_p$. As we did for the groups J and j we construct an isomorphism $\varepsilon: F \simeq f$. Put $C(x, y) = \tau(\varepsilon^{-1}(c(x, y)))$ for any $x, y \in J$. A simple reformulation of Lemma 3.17 in terms of the mapping $C: J \times J \to \mathbb{F}_p$ leads to the following result.

LEMMA 3.19. *Suppose j is a Jordan subgroup of the Lie group G and $C(x, y): J \times J \to \mathbb{F}_p$ is the mapping constructed above. Then*

(1) *$C(x, y)$ is a skew-symmetric bilinear form on J*;
(2) *the Weyl J-group W preserves the form $C(x, y)$*;
(3) *under any other identification τ' of the one-dimensional vector space F and the field $\mathbb{F}_p$ the form $C(x, y)$ is replaced by a proportional one*: $C'(x, y) = \mu C(x, y)$, $\mu \in \mathbb{F}_p^*$.

The role of the form $C(x, y)$ is evident from the following proposition.

PROPOSITION 3.20. *Suppose $\langle x, y\rangle$ is any bilinear form on the vector space J associated with the Jordan subgroup j of G that is invariant under the subgroup W_t of the Weyl J-group W. Then it agrees, to within a factor of proportionality, with the form $C(x, y)$: $\langle x, y\rangle = \lambda C(x, y)$, $\lambda \in \mathbb{F}_p$.*

PROOF. Suppose the bilinear form $C(x, y)$ on J constructed above is not identically zero. Then it is proportional to any other W_t-invariant bilinear form, since, by what was proved earlier, W_t is irreducible. Now suppose $C(x, y)$ is the zero form. We must show that in this case the group W_t does not preserve any nonzero bilinear form on J. Assume there exists a bilinear form $\langle x, y\rangle \not\equiv 0$ that is invariant under W_t. Then, by what was proved in §2, for each J-root α the subspace $J_{(\alpha)}$ is one-dimensional. Therefore each subgroup $j_{(\alpha)} = j \cap G_{(\alpha)}$ of j is cyclic: $j_{(\alpha)} \simeq \mathbb{Z}_p$. We will prove that j is not entirely contained in the subgroup $H = Z_G(j_{(\alpha)})^0$. Let $\mathscr{L}$ be the subalgebra of the complex Lie algebra $\mathfrak{J}$ of G corresponding to H. Obviously $\mathscr{L} = \Sigma_{\beta(J_{(\alpha)})=0}\mathfrak{J}_\beta$. But since, by assumption, the group W_t preserves the nondegenerate bilinear form $\langle x, y\rangle$, it follows that $\beta(J_{(\alpha)}) = 0$ is equivalent to the condition $\langle \alpha, \beta\rangle = 0$, or $\alpha(J_{(\beta)}) = 0$. But then, by Proposition 3.13, $\beta(J_{(\alpha)}) = 0$ implies $[\mathfrak{J}_\alpha, \mathfrak{J}_\beta] = 0$. Thus we have shown that the root subgroup $G_{(\alpha)}$ commutes with H. It is easy to see that $G_{(\alpha)} = Z_G(H)^0$. Now suppose the subgroup $j_{(\beta)}$ is contained in H for some J-root β. Then obviously $Z_G(H)^0 \subset Z_G(j_{(\beta)})^0$ and, since $G_{(\alpha)} = Z_G(H)^0$, we have $\alpha(J_{(\beta)}) = 0$. Thus $\langle \alpha, \beta\rangle = 0$. Therefore H contains only those subgroups $j_{(\beta)}$ for which $\langle \alpha, \beta\rangle = 0$. Since the space J^* is generated by the J-root system Ω, there exists a J-root β such that $\langle \alpha, \beta\rangle \neq 0$. Thus $j \not\subset H$. But since the subgroup $j_{(\alpha)}$ is cyclic, it follows that in the simply connected covering group $\widehat{G}$ of G the centralizer $Z_{\widehat{G}}(\hat{j}_{(\alpha)})$ of the preimage $\hat{j}_{(\alpha)}$ of the group $j_{(\alpha)}$ is connected. Therefore, $C(a, x) = 0$ if and only if $\varepsilon(x) \in H$. It follows from what has been proved that the form $C(x, y)$ is nonzero, which contradicts our assumption. Thus our assumption of the existence of a nonzero W_t-invariant bilinear form has led to a contradiction. The proposition is proved.

We will now complete the proof of Proposition 3.16 for nonclassical Jordan subgroups.

It follows from the W-invariance of the form $C(x, y)$ and Proposition 3.20 that if the Weyl J-group W does not preserve any nonzero bilinear form, then the same is true of the group W_t. But then W_t coincides with the special linear group, $W_t = \mathrm{SL}(J)$ (by Proposition 2.4). Thus, by Lemma 2.5, the J-root system Ω is equal to $J^*\backslash\{0\}$, and for each $\alpha \in \Omega$ we have $J_{(\alpha)} = \operatorname{Ker}\alpha$. Passing to subgroups of j, we obtain $k_{(\alpha)} = \varepsilon(\operatorname{Ker}\alpha) = j_{(\alpha)}$. But since, by definition, $G_{(\alpha)} = Z_G(k_{(\alpha)})^0$, and $G_{(\alpha)}$ is a torus, it is clear that $G_{(\alpha)}$ is a maximal torus of G. Now assume the Weyl J-group W is not equal to its subgroup $W_t = \mathrm{SL}(J)$. Then clearly W contains a diagonalizable element σ that acts identically on the kernel $\operatorname{Ker}\alpha$ of some J-root $\alpha \in \Omega$. The order q of this element σ is obviously relatively prime to p, $(p, q) = 1$. Choose a preimage $\overline{\sigma}$ of this element σ under the homomorphism $A\colon N_G(j) \to W$ so that the order of $\overline{\sigma}$ will also be equal to q. The element $\overline{\sigma}$, by construction, commutes with the subgroup $k_{(\alpha)} = \varepsilon(\operatorname{Ker}\alpha)$ of j, i.e., $\overline{\sigma} \in Z_G(k_{(\alpha)})$. By Lemma 3.10, $|Z_G(k_{(\alpha)})/Z_G(k_{(\alpha)})^0| = p^m$. Thus $\overline{\sigma} \in Z_G(k_{(\alpha)})^0$. The group $Z_G(k_{(\alpha)})^0$ is, by definition, a root subgroup, $G_{(\alpha)} = Z_G(k_{(\alpha)})^0$. Suppose $a \in j\backslash k_{(\alpha)}$. Then it is clear that conjugation of the element a by $\overline{\sigma}$ leaves it in the same coset of $Z_G(k_{(\alpha)})$ relative to the subgroup $Z_G(k_{(\alpha)})^0$. But this means that $\overline{\sigma} a \overline{\sigma}^{-1} = a \pmod{k_{(\alpha)}}$. This contradicts the diagonalizability of the element $\sigma = A(\overline{\sigma})$. Thus our original assumption is false and $W = W_t$. Proposition 3.16 is completely proved.

Thus in the course of studying the structure of the group G and a Jordan subgroup we have proved Theorem 1.4 and its complement, Proposition 1.6, which were stated in §1. Indeed, they follow automatically from Propositions 3.16 and 3.11. Finally, we will prove two more propositions, which will be used in subsequent sections of this paper.

PROPOSITION 3.21. *Suppose j is a Jordan subgroup of the Lie group G. Then j is a classical Jordan subgroup if and only if $\dim_{\mathbb{C}} \mathfrak{J}_\alpha = 1$ for each J-root α.*

PROOF. Suppose $\dim_{\mathbb{C}} \mathfrak{J}_\alpha = 1$ for each J-root α. Then it follows from Corollary 3.5 that $\dim_{\mathbb{C}} G_{(\alpha)} = p - 1$ and $z_{(\alpha)} \simeq \mathbb{Z}_p$, where $z_{(\alpha)} = Z_G(j) \cap G_{(\alpha)}$. Since $j_{(\alpha)}$ is a nonidentity subgroup of $z_{(\alpha)}$, it follows that $j_{(\alpha)} = z_{(\alpha)}$. Thus the subspace $J_{(\alpha)}$ corresponding to $j_{(\alpha)}$ in the vector space J associated with the Jordan subgroup j is one-dimensional: $\dim_{\mathbb{F}_p} J_{(\alpha)} = 1$ for each $\alpha \in \Omega$. But then it follows from what was proved in §2 that the Weyl J-group W preserves some nonzero bilinear form on J, i.e., j is a classical Jordan subgroup.

Conversely, suppose we know that the Weyl J-group W preserves a nonzero bilinear form $\langle x, y\rangle$ on J. By Proposition 3.20, this form is proportional to $C(x, y)$. By what was proved in §2, $\dim_{\mathbb{F}_p} J_{(\alpha)} = 1$ for each J-root α. Since, by Lemma 2.14, $J_{(\alpha)} \neq J_{(\beta)}$ if $\alpha \neq \lambda\beta$, it follows that $J_{(\alpha)} \cap J_{(\beta)} = 0$ for any $\alpha, \beta \in \Omega$, $\alpha \neq \lambda\beta$. As we did earlier, we denote by z the group $z = Z_G(j)$, and by $z_{(\alpha)}$ its intersection with the root subgroup $G_{(\alpha)}$: $z_{(\alpha)} = z \cap G_{(\alpha)}$. Let $M_{(\alpha)} = \{x \in z_{(\alpha)} | gxg^{-1} \equiv x \pmod{j_{(\alpha)}}\ \forall g \in z\}$. We want to prove that $z_{(\alpha)} = j_{(\alpha)}$ for each J-root α. Assume this is not so, i.e., $z_{(\alpha)} \neq j_{(\alpha)}$ for some $\alpha \in \Omega$. Then $M_{(\alpha)} \neq j_{(\alpha)}$, since the subgroup

$M_{(\alpha)}/j_{(\alpha)}$ coincides with the set of fixed points of the action of the p-group z on the group $z_{(\alpha)}/j_{(\alpha)}$, whose order is a power of p. Now suppose $a \in M_{(\alpha)}$ and $b \in M_{(\beta)}$. It follows from the definition of the subgroup $M_{(\alpha)}$ that the element $aba^{-1}b^{-1}$ lies in the intersection $j_{(\alpha)} \cap j_{(\beta)}$, $aba^{-1}b^{-1} \in j_{(\alpha)} \cap j_{(\beta)}$. But, by what has been proved, $j_{(\alpha)} \cap j_{(\beta)} = \{e\}$, hence $aba^{-1}b^{-1} = e$. We have proved that the subgroup $M = \{M_{(\alpha)}\}$ of z generated by all of the $M_{(\alpha)}$ is commutative. It is clear that $M \supset j$. By Corollary 3.6, there exists an element $r \in G_{(\alpha)}$ such that $rar^{-1} = am$, where $a \in j \backslash k_{(\alpha)}$ and $m \in M_{(\alpha)} \backslash j_{(\alpha)}$. Then $r \in N_G(z)$ and, by Lemma 3.8, $r \in N_G(j)$. Thus $m \in j$ and we have arrived at a contradiction. Therefore our original assumption is false, i.e., $z_{(\alpha)} = j_{(\alpha)}$ for each α. But since $\dim_{F_p} J_{(\alpha)} = 1$, hence $j_{(\alpha)} \simeq \mathbb{Z}_p$ and $z_{(\alpha)} \simeq \mathbb{Z}_p$, it follows from Corollary 3.5 that $\dim_{\mathbb{C}} \mathfrak{J}_\alpha = 1$ for each $\alpha \in \Omega$. The proposition is proved.

To find all nonclassical Jordan subgroups (§5) we need

PROPOSITION 3.22. *Suppose j is a Jordan subgroup of the Lie group G that is not a classical Jordan subgroup. Then*

(1) *the Weyl J-group W coincides with the special linear group*: $W = \mathrm{SL}(J)$;

(2) *for each J-root α the root subgroup $G_{(\alpha)}$ is a maximal torus of G*;

(3) *if $n = \dim_{\mathbb{F}_p} J$, there exists a natural number $m > 1$ such that* $\operatorname{rank} G = m(p-1)$ *and* $\dim_{\mathbb{C}} G = m(p^n - 1)$.

PROOF. Assertion (1) has already been proved above. Assertion (2) can be proved as follows. By Lemma 2.5, for any $\alpha \in \Omega$ we have $k_{(\alpha)} = j_{(\alpha)}$, where $j_{(\alpha)} = j \cap G_{(\alpha)}$ and $k_{(\alpha)} = \varepsilon(\operatorname{Ker} \alpha)$. Thus $k_{(\alpha)} \subset G_{(\alpha)}$ and, by definition, $G_{(\alpha)} = Z_G(k_{(\alpha)})^0$. Since $G_{(\alpha)}$ is a torus, it is clear that $G_{(\alpha)}$ is a maximal torus of G.

Let us prove (3). Suppose $m = \dim_{\mathbb{C}} \mathfrak{J}_\alpha$ for some J-root $\alpha \in \Omega$. Since the Weyl J-group W acts transitively on the set of J-roots (Proposition 2.16), it follows that $\dim_{\mathbb{C}} \mathfrak{J}_\beta = m$ for each J-root $\beta \in \Omega$. Since the J-root system Ω is equal to $J^* \backslash \{0\}$ (Lemma 2.5), we have $\dim_{\mathbb{C}} \mathfrak{J} = m(p^n - 1)$. Since, as was proved above, $G_{(\alpha)}$ is a maximal torus of G, it follows that $\operatorname{rank} G = \dim_{\mathbb{C}} G_{(\alpha)}$. But $\mathfrak{J}_{(\alpha)} = \dot{\Sigma}_{\lambda \in F_p} \mathfrak{J}_{\lambda\alpha}$, hence $\dim_{\mathbb{C}} G_{(\alpha)} = m(p-1)$. Therefore $\operatorname{rank} G = m(p-1)$. Since it was given that j is not a classical Jordan subgroup of G, it follows from Proposition 3.21 that $m > 1$. The proposition is proved.

§4. Existence and uniqueness theorems for classical Jordan subgroups

In this section we will prove existence and uniqueness theorems for a classical Jordan subgroup j of a group G with an abstractly defined Weyl J-group

W. These theorems together with a description of all abstract Weyl J-groups, i.e., irreducible linear groups over $\mathbb{F}_p$ generated by transvections, provide a complete classification of classical Jordan subgroups. The final results are exhibited in Table 1. A description of all irreducible linear groups generated by transvections was given in [9]. The statement of this result, suited to the situation studied in the present paper, can be found in §2 (Theorem 2.10).

We will first prove a theorem on the existence of a group G and classical Jordan subgroup j of G with a prescribed abstract Weyl J-group W of classical type (Theorem 1.11).

So suppose there is given a linear group W generated by transvections, acting in a vector space J over the field $\mathbb{F}_p$ and preserving a nondegenerate bilinear form $\langle x, y\rangle$ on J. By Proposition 2.6, the form $\langle x, y\rangle$ is skew-symmetric. The group W is obviously contained in the symplectic group $\overline{W} = \mathrm{Sp}(J)$ of J relative to the skew-symmetric bilinear form $\langle x, y\rangle$: $W \subset \overline{W}$. The symplectic group $\overline{W}$ is also an irreducible group generated by transvections (see Theorem 2.10). It is known that the symplectic group is completely determined by the field $\mathbb{F}_p$, i.e. by the prime number $p = \operatorname{char}\mathbb{F}_p$, and the dimension $n = \dim_{\mathbb{F}_p} J$ of the vector space J, which must be an even number: $n \equiv 0 \bmod 2$. But for each prime p and even number n the existence theorem for a Jordan subgroup j of a group $\overline{G}$ such that the Weyl J-group $W(\overline{G}, j)$ is the same as $\overline{W} = \mathrm{Sp}(J)$, where $n = \dim_{\mathbb{F}_p} J$, was proved by presentation of an example (Proposition 1.2).

Suppose j is a classical Jordan subgroup of a group $\overline{G}$ such that $W(\overline{G}, j) = \overline{W}$. We will construct a connected simple subgroup G of $\overline{G}$ containing j, $G \supset j$, in which j is a classical Jordan subgroup such that $W(G, j) = W$, where W is the prescribed group. Consider the grading $\overline{\mathfrak{J}} = \dot{\Sigma}_{\alpha\in\overline{\Omega}} \mathfrak{J}_\alpha$ of the complex Lie algebra $\overline{\mathfrak{J}}$ of G associated with the Jordan subgroup j. The J-root system $\overline{\Omega}$ of $\overline{G}$ relative to j obviously contains the T-root system Ω of W, $\overline{\Omega} \supset \Omega$. Put $\mathfrak{J} = \dot{\Sigma}_{\alpha\in\Omega} \mathfrak{J}_\alpha$. We will prove that the subspace $\mathfrak{J}$ is a subalgebra of the Lie algebra $\overline{\mathfrak{J}}$. Indeed, if $\alpha, \beta \in \Omega$ and $[\mathfrak{J}_\alpha, \mathfrak{J}_\beta] \neq 0$, then it follows from Proposition 3.13 that $\langle \alpha, \beta\rangle \neq 0$. But then, by Proposition 2.7, $\alpha + \beta \in \Omega$. Since $[\mathfrak{J}_\alpha, \mathfrak{J}_\beta] \subset \mathfrak{J}_{\alpha+\beta}$, we have proved that $[\mathfrak{J}_\alpha, \mathfrak{J}_\beta] \subset \mathfrak{J}$. Thus $\mathfrak{J}$ is a subalgebra of the Lie algebra $\overline{\mathfrak{J}}$. It is clear that to the subalgebra $\mathfrak{J}$ corresponds the subgroup G of $\overline{G}$ generated by all root subgroups $G_{(\alpha)}$ with $\alpha \in \Omega$, $G = \{G_{(\alpha)} | \alpha \in \Omega\}$. We will show that $j \subset G$. Since the set Ω of T-roots generates the entire dual vector space J^*, $J^* = \mathbb{F}_p[\Omega]$, it follows from the isomorphism $J^* \simeq J$ defined by the form $\langle x, y\rangle$ that the group j is generated by the subgroups $j_{(\alpha)} = j \cap G_{(\alpha)}$, $\alpha \in \Omega$. Since $j_{(\alpha)} \subset G$ for each $\alpha \in \Omega$, we have $j \subset G$. We will now show that W is contained in the image of $N_G(j)$ under the natural epimorphism $A\colon N_{\overline{G}}(j) \to \overline{W}$. Indeed, W is generated by the transvections of the form $r_{\alpha,h}$, $\alpha \in \Omega$, $h \in J_{(\alpha)}$. But it follows from Proposition 3.3 that each such transvection has a preimage $r \in A^{-1}(r_{\alpha,h})$ under the homomorphism $A\colon N_{\overline{G}}(j) \to \overline{W}$ that lies in the root torus $G_{(\alpha)}$, $r \in G_{(\alpha)}$. Since, by construction, $G_{(\alpha)} \subset G$ for each $\alpha \in \Omega$, we have $r \in G$. Consequently, $W \subset A(N_G(j))$. By Proposition 2.12, $W \supset A(N_G(j))$. Thus $A(N_G(j)) = W$. From what has been proved it is trivial to verify that j is a classical Jordan subgroup of G and the Weyl J-group $W(G, j)$ is W, $W(G, j) = W$. The theorem is proved.

We will now prove a uniqueness theorem for classical Jordan subgroups. We will first sharpen the statement of the theorem made in §1 (Theorem 1.12). It is necessary to eliminate the ambiguity in the choice of the bilinear form invariant under the Weyl J-group connected with multiplying the form by a scalar. Note that the invariant form $C(x, y)$ explicitly constructed in §3 from the group G and Jordan subgroup j is also defined only to within a factor of proportionality (see Lemma 3.19). This ambiguity occurs only when $p > 2$, since in the case $p = 2$ each nonzero scalar $\lambda \in \mathbb{F}_2^*$ is equal to unity, $\lambda = 1$. The next lemma enables us to eliminate this ambiguity in the choice of the W-invariant bilinear form. Moreover, this lemma will be used in an essential way in the proof of the uniqueness theorem.

Recall that all nonzero homogeneous subspaces $\mathfrak{J}_\alpha$ in the grading of the Lie algebra $\mathfrak{J}$ of G associated with a classical Jordan subgroup j are one-dimensional (Proposition 3.21).

LEMMA 4.1. *Suppose j is a classical Jordan subgroup of G, its period is $p > 2$, $\mathfrak{J} = \sum^{\boldsymbol{\cdot}}_{\alpha\in\Omega} \mathfrak{J}_\alpha$ is the corresponding grading of the Lie algebra $\mathfrak{J}$ of G, and ε is a fixed primitive* pth *root of unity, $\varepsilon^p = 1$. Then there exist a basis $\{E_\alpha | E_\alpha \in \mathfrak{J}_\alpha,\ \alpha \in \Omega\}$ of the Lie algebra $\mathfrak{J}$ and a bilinear form $\langle x, y\rangle$ on the vector space J associated with j that is invariant under the Weyl J-group W such that $[E_\alpha, E_\beta] = (\varepsilon^{\langle\alpha,\beta\rangle} - \varepsilon^{\langle\beta,\alpha\rangle})E_{\alpha+\beta}$ for any J-roots $\alpha, \beta \in \Omega$. The basis $\{E_\alpha\}$ and bilinear form $\langle x, y\rangle$ are uniquely determined to within a simultaneous change of sign: $E_\alpha \mapsto E'_\alpha = -E_\alpha$ for each $\alpha \in \Omega$ and $\langle x, y\rangle \mapsto \langle x, y\rangle' \equiv -\langle x, y\rangle$.*

The proof of the lemma will be given below.

We now state the theorem:

THEOREM 4.2. *Suppose j' and j are classical Jordan subgroups of connected simple complex algebraic groups G' and G, respectively, J' and J are the corresponding vector spaces, $\langle x, y\rangle'$ and $\langle x, y\rangle$ are bilinear forms on J' and J that are invariant under the Weyl J-groups $W' = W(G', j')$ and $W = W(G, j)$, and, if $p > 2$, these are the forms constructed in Lemma* 4.1. *Suppose also there is given an isomorphism $\theta\colon J' \simeq J$ of the vector spaces J' and J such that $\langle\theta(x), \theta(y)\rangle = \omega\langle x, y\rangle'$, where $\omega = \pm 1$ and θ sends W' into W:*

$$W = \{g \in \mathrm{GL}(J) | g = \theta g' \theta^{-1},\ g' \in W'\}.$$

Then there exists an isomorphism $\Theta\colon G' \simeq G$ of the groups G' and G such *that* $\Theta(j')=j$ *and* $\Theta|_{j'}=\varepsilon\theta\varepsilon^{-1}$, *where* $\varepsilon\colon J\simeq j$ *is the standard isomorphism of the additive group of* J *and the group* j.

PROOF. Note first that from the conditions imposed on the isomorphism θ we obtain $\theta^*(\Omega)=\Omega'=\Omega(G',j')$. We will now prove that constructing the desired isomorphism $\Theta\colon G'\simeq G$ is equivalent to constructing an isomorphism $d\Theta\colon\mathfrak{J}'\simeq\mathfrak{J}$ of the Lie algebras $\mathfrak{J}'$ and $\mathfrak{J}$ of the groups G' and G such that $d\Theta(\mathfrak{J}_{\theta^*(\alpha)})=\mathfrak{J}_\alpha$ for each J-root $\alpha\in\Omega=\Omega(G,j)$ (here $\theta^*\colon J^*\simeq J'^*$ is the isomorphism dual to the given isomorphism θ). Indeed, in our case, prescribing a Lie algebra isomorphism Θ is equivalent to prescribing a Lie group isomorphism, since, by definition, the centers of the Lie groups G' and G are trivial. The conditions imposed on $d\Theta$ are obviously equivalent to the required conditions on Θ, i.e., $\Theta(j')=j$ and $\Theta|_{j'}=\varepsilon\theta\varepsilon^{-1}$.

Thus to prove the theorem it suffices to construct a Lie algebra isomorphism $d\Theta\colon\mathfrak{J}'\simeq\mathfrak{J}$ sending the grading $\mathfrak{J}'=\Sigma^{\cdot}\mathfrak{J}'_\alpha$ of the Lie algebra $\mathfrak{J}'$ into the grading $\mathfrak{J}=\Sigma^{\cdot}\mathfrak{J}_\alpha$ of the Lie algebra $\mathfrak{J}$ in the prescribed (via θ^*) way. We have proved (Proposition 3.21) that for each J-root α relative to a classical Jordan subgroup the subspace $\mathfrak{J}_\alpha$ is one-dimensional: $\dim_{\mathbb{C}}\mathfrak{J}_\alpha=1$. Choose a basis $\{E_\alpha\in\mathfrak{J}_\alpha|\alpha\in\Omega\}$ of the Lie algebra $\mathfrak{J}$. To construct the isomorphism $d\Theta$ we study the structure constants of $\mathfrak{J}$ in this basis. It follows from Proposition 3.13 that

$$[E_\alpha,E_\beta]=\begin{cases}0 & \text{if } \langle\alpha,\beta\rangle=0,\\ \Lambda(\alpha,\beta)E_{\alpha+\beta} & \text{where } \Lambda(\alpha,\beta)\in\mathbb{C}^*,\\ & \text{if } \langle\alpha,\beta\rangle\neq 0.\end{cases}\tag{4.3}$$

Under a change of basis $E_\alpha\mapsto E^1_\alpha=\lambda_\alpha E_\alpha$, $\lambda_\alpha\in\mathbb{C}^*$, the function Λ obviously changes as follows:

$$\Lambda_1(\alpha,\beta)=\lambda_\alpha\lambda_\beta\lambda^{-1}_{\alpha+\beta}\Lambda(\alpha,\beta).\tag{4.4}$$

The expression $\Lambda(\alpha,\beta)$ will be viewed as a function on the set $\Omega\times\Omega$ with values in the field of complex numbers $\mathbb{C}$ by putting $\Lambda(\alpha,\beta)=0$ when $\langle\alpha,\beta\rangle=0$.

The skew-symmetry of the commutator $[X,Y]$ in a Lie algebra and the Jacobi identity are equivalent to the following identities for $\Lambda(\alpha,\beta)$:

$$\Lambda(\alpha,\beta)=-\Lambda(\beta,\alpha),\qquad \alpha,\beta\in\Omega;\tag{4.5}$$

$$\begin{aligned}\Lambda(\alpha,\beta)\Lambda(\alpha+\beta,\gamma)&+\Lambda(\gamma,\alpha)\Lambda(\gamma+\alpha,\beta)\\ &+\Lambda(\beta,\gamma)\Lambda(\beta+\gamma,\alpha)=0,\qquad \alpha,\beta,\gamma\in\Omega.\end{aligned}\tag{4.6}$$

When $p>2$ we can find explicitly all functions $\Lambda(\alpha,\beta)$ satisfying (4.5), (4.6) and vanishing if and only if $\langle\alpha,\beta\rangle=0$, to within a replacement of the form (4.4). This result has already been stated above in Lemma 4.1. Proceeding from this result, we will prove the theorem for $p>2$. Choose a basis $\{E_\alpha|E_\alpha\in\mathfrak{J}_\alpha\}$ of the Lie algebra $\mathfrak{J}$ for which $\Lambda(\alpha,\beta)=\varepsilon^{\langle\alpha,\beta\rangle}-\varepsilon^{\langle\beta,\alpha\rangle}$

(such a basis exists by Lemma 4.1). Choose an analogous basis $\{E'_{\theta^*(\alpha)}\}$ in the Lie algebra $\mathfrak{J}'$. Since the bilinear form $\langle x, y\rangle'$ is determined up to sign (by Lemma 4.1), we may assume the number ω in the statement of the theorem is equal to unity, $\omega = 1$, i.e., $\langle \theta^*(x), \theta^*(y)\rangle' \equiv \langle x, y\rangle$. Thus the structure constants $\Lambda'(\theta^*(\alpha), \theta^*(\beta))$ of the Lie algebra $\mathfrak{J}'$ in the basis $\{E'_{\theta^*(\alpha)}\}$ have the form

$$\Lambda'(\theta^*(\alpha), \theta^*(\beta)) = \varepsilon^{\langle \theta^*(\alpha), \theta^*(\beta)\rangle} - \varepsilon^{\langle \theta^*(\beta), \theta^*(\alpha)\rangle} = \varepsilon^{\langle \alpha, \beta\rangle} - \varepsilon^{\langle \beta, \alpha\rangle}.$$

Now since $\Lambda'(\theta^*(\alpha), \theta^*(\beta)) = \Lambda(\alpha, \beta)$, it is clear that the mapping $d\Theta$: $\mathfrak{J}' \to \mathfrak{J}$ defined in the basis $\{E'_{\theta^*(\alpha)}\}$ by the rule $d\Theta(E'_{\theta^*(\alpha)}) = E_\alpha$ is a Lie algebra isomorphism. The theorem is proved for $p > 2$.

In the case $p = 2$ we will prove uniqueness of the function $\Lambda(\alpha, \beta)$ to within a replacement (4.4). We will use a method similar to the one employed by Tits [14] for making precise the sign of the structure constants $N_{\alpha,\beta}$ of a simple Lie algebra in a Cartan basis. It is obvious that the assertion of the theorem follows from the uniqueness of the function $\Lambda(\alpha, \beta)$, as in the case $p > 2$.

Thus to complete the proof of the theorem we must prove Lemma 4.1 and the following

LEMMA 4.7. *Suppose $\Lambda(\alpha, \beta)$ is a complex-valued function on $\Omega \times \Omega$, where Ω is the J-root system relative to a classical Jordan subgroup of period $p = 2$, satisfying* (4.5), (4.6) *and vanishing if and only if $\langle \alpha, \beta\rangle = 0$. Then this function is uniquely determined to within a replacement* (4.4).

PROOF OF LEMMA 4.1. By the hypothesis of the lemma, $p > 2$. Note that in this case the Weyl J-group W is the symplectic group: $W = \mathrm{Sp}(J)$ (see §2). To prove the lemma we will find all possible functions $\Lambda(\alpha, \beta)$. We will show that we can choose a basis $\{E_\alpha | E_\alpha \in \mathfrak{J}_\alpha\}$ of the Lie algebra $\mathfrak{J}$ such that the function $\Lambda(\alpha, \beta)$ will depend only on the value of the bilinear form $\langle \alpha, \beta\rangle \in \mathbb{F}_p$, i.e., $\Lambda(\alpha, \beta) \equiv \Phi(\langle \alpha, \beta\rangle)$. In turn, this assertion is a consequence of the splittability of the epimorphism A: $N_G(j) \to W$ and certain simple properties of the group $W = \mathrm{Sp}(J)$. Let us proceed with the proof.

PROPOSITION 4.8. *The group $N = N_G(j)$ is a semidirect product $N = W_\tau \cdot Z_G(j)$ of the centralizer $Z_G(j)$ of the Jordan subgroup j and another subgroup W_τ.*

PROOF. To prove the proposition we use the method employed by Wales [16] to study the subgroup $N = N_G(j)$ of the group $G = \mathrm{SL}_p(\mathbb{C})$, p a prime.

Since in our case the group $W = N/Z_G(j)$ is isomorphic to the symplectic group, $W = \mathrm{Sp}(J)$, it follows that the center of W consists of the two elements $\pm e$. Let τ be a preimage of $-e$ in N, $\tau \in N$. Since $Z_G(j)$ is a p-group (by Lemma 3.10), the element τ can be chosen so that $\tau^2 = e$. Put $W_\tau = Z_N(\tau)$.

We will first prove that the group W_τ projects onto the whole group W under the homomorphism $A\colon N \to W$. Suppose $\bar{r}_\alpha$ is any transvection in W and r_α is a preimage of $\bar{r}_\alpha$ in N lying in the root subgroup $G_{(\alpha)}$ (the existence of r_α follows from Propositions 3.11 and 3.3). The commutator $h = \tau r_\alpha \tau^{-1} r_\alpha^{-1}$ obviously belongs to the group $Z_G(j) \cap G_{(\alpha)}$. It follows from Corollary 3.5 and Proposition 3.21 that $Z_G(j) \cap G_{(\alpha)} = j \cap G_{(\alpha)}$. Since $p > 2$, the group $j_{(\alpha)} = j \cap G_{(\alpha)}$ contains an element x such that $x^2 = h$. It can be verified formally that $r_\alpha x$ commutes with τ, hence $r_\alpha x \in W_\tau$. Since W is generated by transvections, W_τ projects onto the whole group W.

It remains to prove that $W_\tau \cap Z_G(j) = \{e\}$. By construction, $\tau a \tau^{-1} = a^{-1}$ for each $a \in j$. Thus $W_\tau \cap j = \{e\}$. Now let A denote the group $W_\tau \cap Z_G(j)$, $A = W_\tau \cap Z_G(j)$. Suppose a is any element of A, $a \in A$. Put $B = Z_N(a)$. It follows easily from the fact that $aG_{(\alpha)}a^{-1} = G_{(\alpha)}$ for each $\alpha \in \Omega$ and from $W_\tau \cap G_{(\alpha)} \cap j = \{e\}$ that $W_\tau \subset B$. It is also clear that $B \supset j$. The representation of the group B by adjoint operators in the space of the Lie algebra $\mathfrak{J}$ of G is irreducible. Indeed, each root subspace $\mathfrak{J}_\alpha$ is one-dimensional, and W (hence also W_τ) acts transitively on the J-root system Ω (Proposition 2.11). But the subalgebra $\mathscr{L}$ of $\mathfrak{J}$ corresponding to the subgroup $H = Z_G(a)^0$ is obviously B-invariant. Thus $\mathscr{L} = \mathfrak{J}$ and, since the center of G is trivial, $a = e$. Therefore $A = \{e\}$, i.e., $W_\tau \cap Z_G(j) = \{e\}$. The proposition is proved.

Now let W_α denote the stationary subgroup of the J-root α in W, i.e., $W_\alpha = \{g \in W \mid g^*\alpha = \alpha\}$. It is clear that the corresponding subgroup W_{τ_α} of W_τ consists of the following transformations: $W_{\tau_\alpha} = \{g \in W_\tau \mid \operatorname{Ad} g(\mathfrak{J}_\alpha) = \mathfrak{J}_\alpha\}$.

LEMMA 4.9. *For each $g \in W_{\tau_\alpha}$ and vector $E_\alpha \in \mathfrak{J}_\alpha \setminus \{0\}$ we have* $\operatorname{Ad} g(E_\alpha) = E_\alpha$.

PROOF. We will first show that the group W_α is generated by its commutator (W_α, W_α) and the transvections lying in W_α. Indeed, since $W = \operatorname{Sp}(J)$, it is easy to find the stationary group of an arbitrary linear form α in the group $\operatorname{Sp}(J)$ and verify the desired result.

For elements of the commutator (W_α, W_α) the assertion of the lemma is obvious, since $\dim_{\mathbb{C}} \mathfrak{J}_\alpha = 1$. Now suppose r_β is an element of W_{τ_α} corresponding to a transvection in W. It was shown above, in the proof of Proposition 4.8, that $r_\beta \in G_{(\beta)}$. Since $r_\beta^*\alpha = \alpha$, it follows from Lemma 2.1 that $\langle \beta, \alpha \rangle = 0$, hence the tori $G_{(\alpha)}$ and $G_{(\beta)}$ commute (Proposition 3.13). Thus $\operatorname{Ad} r_\beta(E_\alpha) = E_\alpha$. The lemma is proved.

Lemma 4.9 trivially implies

COROLLARY 4.10. *There exists a basis $\{E_\alpha \mid E_\alpha \in \mathfrak{J}_\alpha,\ \alpha \in \Omega\}$ such that for each $g \in W_\tau$ we have $\operatorname{Ad} g(E_\alpha) = E_{g^*\alpha}$. If $\{E'_\alpha\}$ is another basis with the same property, then there exists $\lambda \in \mathbb{C}^*$ such that $E'_\alpha = \lambda E_\alpha$ for each $\alpha \in \Omega$.*

Now suppose $\{E_\alpha\}$ is a basis from Corollary 4.10. In this basis the structure constants $\Lambda(\alpha, \beta)$ obviously have the property that $\Lambda(g^*\alpha, g^*\beta) = \Lambda(\alpha, \beta)$ for each $g \in W$. In other words, the function $\Lambda(\alpha, \beta)$ is constant on the orbits of the group $W = \mathrm{Sp}(J)$ under its action on the set $\Omega \times \Omega$. It is well known from the theory of skew-symmetric bilinear forms that the skew-symmetric bilinear form $\langle \alpha, \beta \rangle$ separates the orbits of $\mathrm{Sp}(J)$. Thus there exists a function $\Phi: \mathbb{F}_p \to \mathbb{C}$ such that $\Lambda(\alpha, \beta) = \Phi(\langle \alpha, \beta \rangle)$. Conditions (4.3) and (4.6) are equivalent to the following conditions on the function Φ:

(4.11) $\Phi(0) = 0$, and $\Phi(x) \neq 0$ if $x \neq 0$;

(4.12) $\Phi(-x) = -\Phi(x)$;

(4.13) $\Phi(x)\Phi(y + z) + \Phi(-y)\Phi(-z + x) + \Phi(z)\Phi(-x - y) = 0$ for any $x, y, z \in \mathbb{F}_p$.

It follows from Corollary 4.10 and formula (4.4) for admissable replacements of the function $\Lambda(\alpha, \beta)$ that the function Φ is defined to within a proportionality factor.

LEMMA 4.14. *All solutions of the system of equations* (4.11)–(4.13) *can be obtained by multiplying by an arbitrary scalar* $\lambda \in \mathbb{C}^*$ *each of the* $(p-1)/2$ *functions* $\Phi_l(k) = \varepsilon^{lk} - \varepsilon^{-lk}$, *where* ε *is a fixed primitive* pth *root of unity and* $l = 1, 2, \dots, (p-1)/2$. *The functions* $\Phi_l(k)$ *with different* $l = 1, \dots, (p-1)/2$ *are not proportional.*

PROOF. It can be verified trivially that each function Φ_l is a solution of the system of equations. It is also clear that functions Φ_l with different $l = 1, \dots, (p-1)/2$ are not proportional. We will show that the number of solutions, to within multiplication by a scalar, is at most $(p-1)/2$. Suppose Φ is any solution of the system (4.11)–(4.13). By multiplying by a scalar we can ensure that $\Phi(1) = 1$. Let $\Phi(2) = \mu$. From (4.13), using the substitution $x = y = 1$, we obtain $\Phi(z+1) = \Phi(2)\Phi(z) - \Phi(z-1)$. Since $\Phi(2) = \mu$, it is obvious that $\Phi(z+1) = \mu^z + \cdots$, a polynomial of degree z in μ $(z = 2, 3, \dots, p-1)$. The elements $(p-1)/2$ and $(p+1)/2$ are mutually inverse in the field $\mathbb{F}_p$. Thus it follows from (4.12) that $\Phi((p+1)/2) + \Phi((p-1)/2) = 0$. Consequently, μ satisfies an equation of degree $(p-1)/2$. Since it is clear from what has been proved that the function Φ is uniquely determined by μ, we conclude that the number of solutions of the system of equations (4.11)–(4.13) with $\Phi(1) = 1$ is at most $(p-1)/2$. The lemma is proved.

Thus we have shown that in some basis

$$\Lambda(\alpha, \beta) = \varepsilon^{l\langle \alpha, \beta \rangle} - \varepsilon^{l\langle \beta, \alpha \rangle}. \tag{4.15}$$

We will now renormalize the bilinear form $\langle x, y \rangle$. Put $\langle x, y \rangle' = l\langle x, y \rangle$. Then $\Lambda(\alpha, \beta) = \varepsilon^{\langle \alpha, \beta \rangle'} - \varepsilon^{\langle \beta, \alpha \rangle'}$. It is clear that the only replacement of the basis $\{E_\alpha\}$ and W-invariant bilinear form $\langle x, y \rangle'$ preserving the form of the function $\Lambda(\alpha, \beta)$ is $E_\alpha \mapsto -E_\alpha$ for each $\alpha \in \Omega$ and $\langle x, y \rangle' \mapsto -\langle x, y \rangle'$. Lemma 4.1 is proved.

Let us now turn to the proof of Lemma 4.7. We denote by (X, Y) an $\operatorname{Ad} G$-invariant scalar product on the Lie algebra $\mathfrak{J}$. Using the invariance of (X, Y) and the fact that $-\alpha = \alpha$ in characteristic $p = 2$, we see that the restriction $(X, Y)|_{\mathfrak{J}_\alpha}$ of the form (X, Y) to each root subspace $\mathfrak{J}_\alpha$ is nondegenerate. Therefore we can choose a basis of the Lie algebra $\mathfrak{J}$ in the form $\{E_\alpha \in \mathfrak{J}_\alpha | (E_\alpha, E_\alpha) = 1, \alpha \in \Omega\}$.

LEMMA 4.16. *In the basis* $\{E_\alpha \in \mathfrak{J}_\alpha | (E_\alpha, E_\alpha) = 1, \alpha \in \Omega\}$ *the structure constants* $\Lambda(\alpha, \beta)$ *of the Lie algebra* $\mathfrak{J}$ *have the form* $\Lambda(\alpha, \beta) = \pm\lambda$ *for all* $\alpha, \beta \in \Omega$, $\langle \alpha, \beta \rangle \neq 0$, *where* $\lambda \in \mathbb{C}^*$ *is a scalar.*

PROOF. Since $(\operatorname{Ad} g(E_\alpha), \operatorname{Ad} g(E_\alpha)) = (E_\alpha, E_\alpha)$ for each $g \in N_G(j)$, it is obvious that $\operatorname{Ad} g(E_\alpha) = \pm E_{g^*\alpha}$. The lemma now follows from the transitivity of the action of W on the set of pairs $\{(\alpha, \beta) | \alpha, \beta \in \Omega, \langle \alpha, \beta \rangle = 1\}$ (Lemma 2.11).

The lemma trivially implies

COROLLARY 4.17. *We can normalize the invariant scalar product* (X, Y) *on the Lie algebra* $\mathfrak{J}$ *so that in a basis* $\{E_\alpha \in \mathfrak{J}_\alpha | (E_\alpha, E_\alpha) = 1\}$ *the structure constants have the form* $\Lambda(\alpha, \beta) = \pm 1$ *for all* $\alpha, \beta \in \Omega$, $\langle \alpha, \beta \rangle \neq 0$.

We will assume below that the invariant scalar product has been chosen as in Corollary 4.17.

Now consider the intersection $R_{(\alpha)} = G_{(\alpha)} \cap N_G(j)$ of the one-dimensional root torus $G_{(\alpha)}$ and the subgroup $N = N_G(j)$. It is easy to show, using Corollary 3.5, that the group $R_{(\alpha)}$ is isomorphic to $\mathbb{Z}_4$, $R_{(\alpha)} \simeq \mathbb{Z}_4$, and $R_{(\alpha)} \cap j = G_{(\alpha)} \cap j = \mathbb{Z}_2$. Both elements of $R_{(\alpha)} \backslash j_{(\alpha)}$ project, obviously, into the transvection $\bar{r}_\alpha$ under the homomorphism $A: N_G(j) \to W$. Now, using the elements of $R_{(\alpha)} \backslash j_{(\alpha)}$ for each J-root $\alpha \in \Omega$, we can parametrize the ambiguity in the choice of basis $\{E_\alpha\}$ in Corollary 4.17 connected with the possibility of replacing E_α by $E'_\alpha = -E_\alpha$ for each $\alpha \in \Omega$ separately.

Suppose $\bar{\mathfrak{J}}_\alpha$ is the real part of the subspace $\mathfrak{J}_\alpha$ and $\exp: \bar{\mathfrak{J}}_\alpha \to G_{(\alpha)}$ is the exponential mapping. Suppose $r_\alpha \in R_{(\alpha)} \backslash j_{(\alpha)}$. Since $G_{(\alpha)}$ is a one-dimensional torus and r_α is an element of order 4 in it, it is easy to show that the preimage $\exp^{-1}(r_\alpha)$ contains exactly one vector $E'(r_\alpha) \in \bar{\mathfrak{J}}_\alpha$ of smallest length relative to the scalar product (E'_α, E'_α). We now put $E(r_\alpha) = E'(r_\alpha)/(E'(r_\alpha), E'(r_\alpha))$.

LEMMA 4.18. *Suppose* r'_α *is an element of* $R_{(\alpha)} \backslash j_{(\alpha)}$ *different from* r_α, *i.e.,* $r'_\alpha = r_\alpha \cdot h_\alpha$, *where* $h_\alpha \in j_{(\alpha)} \backslash \{e\}$. *Then* $E(r_\alpha h_\alpha) = -E(r_\alpha)$.

The proof is trivial.

Now let $M = \bigcup_{\alpha \in \Omega} (R_{(\alpha)} \backslash j_{(\alpha)})$. In the direct product $M \times M \times M$ of three copies of M we distinguish the subset

$$\Delta = \{(r_\alpha, r_\beta, r_\gamma | r_x \in R_{(x)} \backslash j(x),\ x = \alpha, \beta, \gamma;\ \alpha + \beta + \gamma = 0; \langle \alpha, \beta \rangle \neq 0\}.$$

We define a complex-valued function $\delta = \delta(r_\alpha, r_\beta, r_\gamma)$ on the set Δ by the formula

$$[E(r_\alpha), E(r_\beta)] = \delta(r_\alpha, r_\beta, r_\gamma)E(r_\gamma), \tag{4.19}$$

where $\gamma = \alpha + \beta$.

LEMMA 4.20. *The values of the function* δ *defined by* (4.19) *are* ± 1, *i.e.,* $\delta(r_\alpha, r_\beta, r_\gamma) \in \{\pm 1\}$. *If* $h_\alpha \in j_{(\alpha)} \backslash \{e\}$, $h_\beta \in j_{(\beta)} \backslash \{e\}$, $h_\gamma \in j_{(\gamma)} \backslash \{e\}$, *then*

$$\delta(r_\alpha h_\alpha, r_\beta, r_\gamma) = \delta(r_\alpha, r_\beta h_\beta, r_\gamma) = \delta(r_\alpha, r_\beta, r_\gamma h_\gamma) = -\delta(r_\alpha, r_\beta, r_\gamma).$$

The proof of the lemma follows trivially from Corollary 4.17 and Lemma 4.18.

Now suppose $r_\alpha \in R_{(\alpha)} \backslash j_{(\alpha)}$, $r_\beta \in R_{(\beta)} \backslash j_{(\beta)}$, and $\langle \alpha, \beta \rangle \neq 0$. Then the element $r_\alpha r_\beta r_\alpha^{-1}$ obviously lies in $R_{(\alpha+\beta)} \backslash j_{(\alpha+\beta)}$.

LEMMA 4.21. *For any* $\alpha, \beta \in \Omega$ *such that* $\langle \alpha, \beta \rangle \neq 0$ *and any* $r_\alpha \in R_{(\alpha)} \backslash j_{(\alpha)}$, $r_\beta \in R_{(\beta)} \backslash j_{(\beta)}$ *we have*

$$\delta(r_\alpha, r_\beta, r_\alpha r_\beta r_\alpha^{-1}) = 1.$$

PROOF. The subgroup $G_{(\alpha,\beta)}$ of G generated by the tori $G_{(\alpha)}$, $G_{(\beta)}$, $G_{(\alpha+\beta)}$ is isomorphic to SO_3, $G_{(\alpha,\beta)} \simeq SO_3$. Indeed, its Lie algebra $\mathfrak{J}_{(\alpha,\beta)} = \mathfrak{J}_\alpha \dot{+} \mathfrak{J}_\beta \dot{+} \mathfrak{J}_{\alpha+\beta}$ is three-dimensional and noncommutative, hence $\mathfrak{J}_{(\alpha,\beta)} \simeq \mathfrak{so}(3)$. Moreover, it is easy to verify that $G_{(\alpha,\beta)} \simeq SL_2$. Thus $G_{(\alpha,\beta)} \simeq SO_3$. In some basis the subgroups of G in which we are interested can be described by the following matrices:

$$j \cap G_{(\alpha,\beta)} = \left\{ \begin{pmatrix} \varepsilon_1 & & \\ & \varepsilon_2 & \\ & & \varepsilon_3 \end{pmatrix} \middle| \varepsilon_i = \pm 1, \ \varepsilon_1\varepsilon_2\varepsilon_3 = 1 \right\};$$

$$j_{(\alpha)} = \left\{ e, \begin{pmatrix} -1 & & \\ & -1 & \\ & & -1 \end{pmatrix} \right\}; \qquad j_{(\beta)} = \left\{ e, \begin{pmatrix} -1 & & \\ & -1 & \\ & & -1 \end{pmatrix} \right\};$$

$$j_{(\alpha+\beta)} = \left\{ e, \begin{pmatrix} -1 & & \\ & 1 & \\ & & -1 \end{pmatrix} \right\};$$

$$r_\alpha = \begin{pmatrix} 0 & -1 & 0 \\ 1 & 0 & 0 \\ 0 & 0 & 1 \end{pmatrix}; \qquad r_\beta = \begin{pmatrix} 1 & 0 & 0 \\ 0 & 0 & -1 \\ 0 & 1 & 0 \end{pmatrix}.$$

The lemma can now be proved by direct calculation.

The properties of the function δ proved in Lemmas 4.20 and 4.21 obviously define it uniquely. It is clear that prescribing the function δ is equivalent to prescribing the function $\Lambda(\alpha, \beta)$. Lemma 4.7 is proved.

Thus we have proved existence and uniqueness theorems for a classical Jordan subgroup with a prescribed abstract Weyl J-group. We will now deduce from these theorems the classification theorem, Theorem 1.13, stated in §1.

The existence and uniqueness theorems establish a one-to-one correspondence between the irreducible groups generated by transvections over the fields $\mathbb{F}_p$ preserving a nonzero bilinear form and the classical Jordan subgroups to within an automorphism of G. Consequently, Theorem 1.13 follows from the classification of linear groups generated by transvections (Theorem 2.10). The only thing that remains is to determine to which Lie algebra $\mathfrak{I}$ corresponds the given group G generated by transvections. This is easily done, starting from the representation of the Lie algebra $\mathfrak{I}$ of G constructed in the proof of the existence theorem. The theorem is proved.

§5. Description of the nonclassical Jordan subgroups

In this section we will complete the description of all Jordan subgroups of simple complex algebraic groups. Since the classical Jordan subgroups have already been described in §4, it remains only to describe the nonclassical Jordan subgroups, i.e., those Jordan subgroups j whose Weyl J-group W does not preserve any nonzero bilinear form. An important role in this section is played by Proposition 3.22, which describes the Weyl J-group W of G relative to a nonclassical Jordan subgroup j and properties of the grading $\mathfrak{I} = \dot{\Sigma}_{\alpha\in\Omega}\mathfrak{I}_\alpha$ of the Lie algebra $\mathfrak{I}$ of G associated with j.

We will first prove that nonclassical Jordan subgroups can exist only in exceptional simple algebraic groups.

Proposition 5.1. *Suppose G is a classical simple algebraic group (i.e., G is of type A_r, B_r, C_r, or D_r). Then G has no nonclassical Jordan subgroups.*

Proof. Suppose G is a simple group and j is a nonclassical Jordan subgroup. Consider a universal covering group $\widehat{G}$ of G. Let $\hat{j}$ be the complete preimage of j in $\widehat{G}$. Then $\hat{j}$ is a commutative subgroup of $\widehat{G}$ (Proposition 3.22). From this we can prove Proposition 5.1 for groups G of type A_r or C_r. Indeed, it is known that each commutative subgroup of SL_{r+1} or Sp_{2r} lies in some torus (see, e.g., [7]). Consequently, the centralizer $Z_G(j)$ of j contains some torus, which contradicts the definition of a Jordan subgroup.

Thus it remains to consider groups G of type B_r or D_r. We will need to introduce a subgroup $\tilde{z}$ of the centralizer $z = Z_G(j)$ of the Jordan subgroup j. Suppose $\pi\colon \widehat{G} \to G$ is the natural projection of the universal covering group $\widehat{G}$ onto G. Let $\tilde{z} = \pi(M)$, where $M = Z_{\widehat{G}}(\hat{j})$. It is clear that $j \subset \tilde{z}$. Moreover, $\tilde{z}$ is a normal subgroup of the normalizer $N_G(j)$ of the Jordan subgroup j. Starting from properties of a group G of classical type, we will first prove that $\tilde{z}$ is a commutative group (Corollary 5.4 below). In view of the maximality condition on $N_G(j)$ in the definition of a Jordan subgroup, it is easy to deduce that, roughly speaking, $\tilde{z} = j$ (Lemma 5.4). On the other hand, it is easy to show that $\tilde{z}$ almost coincides with the centralizer z of the Jordan subgroup j (Lemma 5.5). From precise formulations of these assertions and Proposition 3.20 we can prove an inequality between the rank r of G and the dimension $n = \dim_{\mathbb{F}_p} J$ of J over $\mathbb{F}_p$, i.e., $r \leq n+1$. Using the formulas of Proposition 3.20, we obtain $r \approx 2^{n-1}$. It follows that the only possibility for the constants r and n is $r = 4$, $n = 3$, and G is of type D_4. It is easy to see that D_4 contains no nonclassical Jordan subgroups.

We will now state and prove the intermediate results.

LEMMA 5.2. *Suppose j is a nonclassical Jordan subgroup of G and $\tilde{z} = \pi(Z_{\widehat{G}}(\pi^{-1}(j)))$, where $\pi: \widehat{G} \to G$ is a universal covering. Then each element $x \in \tilde{z}$ of the subgroup $\tilde{z}$ lies in the center of some semisimple subgroup H of G of maximal rank.*

PROOF. Let $H = Z_G(x)^0$, where $x \in \tilde{z}$. Since the centralizer of an element $\hat{x} \in \pi^{-1}(x)$ in the simply connected, simple algebraic group $\widehat{G}$ is connected (see [12]), it follows that $\hat{j} \subset \widehat{H}$, where $\hat{j} = \pi^{-1}(j)$, $\widehat{H} = \pi^{-1}(H)$. Thus $j \subset H$ and since, by definition, $Z_G(j)^0 = \{e\}$, the center of H is finite. Thus H is a semisimple group. It is clear that H is a subgroup of G of maximal rank and that x lies in the center of H. The lemma is proved.

We will now deduce a corollary from Lemma 5.2 for a group G of type B_r or D_r.

COROLLARY 5.3. *Suppose G is a group of type B_r or D_r and j is a nonclassical Jordan subgroup of G. Then the subgroup $\tilde{z}$ is isomorphic to a direct product of groups $\mathbb{Z}_2$: $\tilde{z} \simeq \mathbb{Z}_2 \times \cdots \times \mathbb{Z}_2$.*

It follows from the classification of semisimple subgroups H of maximal rank and their centers in groups G of type B_r or D_r that the center of each subgroup H is a group of period 2. By Lemma 5.2, $\tilde{z}$ is also a group of period 2. But every group of period 2 is commutative. Thus $\tilde{z} \simeq \mathbb{Z}_2 \times \cdots \times \mathbb{Z}_2$. The corollary is proved.

LEMMA 5.4. *Under the conditions of Corollary 5.3, for each J-root α the intersection $\tilde{z}_{(\alpha)} = \tilde{z} \cap G_{(\alpha)}$ of the subgroup $\tilde{z}$ with the root subgroup $G_{(\alpha)}$ coincides with the subgroup $j_{(\alpha)} = j \cap G_{(\alpha)}$, i.e., $\tilde{z}_{(\alpha)} = j_{(\alpha)}$.*

PROOF. Consider the subgroup A of $\tilde{z}$ generated by all of the $\tilde{z}_{(\alpha)}$, i.e., $A = \{\tilde{z}_{(\alpha)} | \alpha \in \Omega\}$. It is clear that $N_G(A) \supset N_G(j)$. Moreover, it follows easily from the definition of Jordan subgroup that $N_G(A) = N_G(j)$. Let us now assume that $\tilde{z}_{(\alpha)} \neq j_{(\alpha)}$ for some α. Then, by Proposition 3.3, for any $a \in \tilde{z}_{(\alpha)} \backslash j_{(\alpha)}$ there exists an element $r \in N_G(A)$ such that $r a_\alpha r^{-1} = a_\alpha a$, where $a_\alpha \in j \backslash j_{(\alpha)}$ (apply Proposition 3.3 to the subgroup A and root subgroups $G_{(\omega)}$ relative to A). But it is easy to show that $\tilde{z}_{(\alpha)} = \prod_{\omega \in T} A_{(\omega)}$, where $A_{(\omega)} = A \cap G_{(\omega)}$, and $G_{(\alpha)} = \prod_{\omega \in T} G_{(\omega)}$. Since, by what has been proved, $N_G(A) = N_G(j)$, it follows that $r \in N_G(j)$, hence $a \in j$. This contradicts our assumption $a \notin j_{(\alpha)}$. Thus $\tilde{z}_{(\alpha)} = j_{(\alpha)}$ for each α. The lemma is proved.

We will prove one more lemma.

LEMMA 5.5. *Under the conditions of Corollary* 5.3, *suppose* α *is a* J*-root and* $z_{(\alpha)} = Z_G(j) \cap G_{(\alpha)}$, $j_{(\alpha)} = j \cap G_{(\alpha)}$, *where* $G_{(\alpha)}$ *is the root subgroup. Then* $|z_{(\alpha)}/j_{(\alpha)}| \leq 4$.

PROOF. Let Δ denote the center of the universal covering group $\widehat{G}$ of G. Consider the mapping $c\colon z_{(\alpha)} \to \mathrm{Hom}(j, \Delta)$ defined by $c(x)(a) = \hat{x}\hat{a}\hat{x}^{-1}\hat{a}^{-1} \in \Delta$, where $\hat{a}$ and $\hat{x}$ are preimages of $a \in j_{(\alpha)}$ and $x \in z_{(\alpha)}$ in $\widehat{G}$. It is easy to show that the mapping c is correctly defined and is a group homomorphism. It is also clear that for each x the homomorphism $c(x)$ sends $j_{(\alpha)}$ into $\{e\}$, i.e., $c(x)(j_{(\alpha)}) = \{e\}$, and that the kernel of the homomorphism c is $\widetilde{z}_{(\alpha)}$, $\operatorname{Ker} c = \widetilde{z}_{(\alpha)}$. By Lemma 5.4, $\widetilde{z}_{(\alpha)} = j_{(\alpha)}$, hence we obtain an embedding of the group $z_{(\alpha)}/j_{(\alpha)}$ into the group $\mathrm{Hom}(j/j_{(\alpha)}, \Delta)$. For nonclassical Jordan subgroups, $j/j_{(\alpha)} = \mathbb{Z}_p$ (by Proposition 3.20). In our case, $p = 2$. Moreover, $\Delta = \mathbb{Z}_2, \mathbb{Z}_4, \mathbb{Z}_2 \times \mathbb{Z}_2$ for groups of type B_r, D_{2m+1}, D_{2m}, respectively. Thus $|z_{(\alpha)}/j_{(\alpha)}| \leq 4$, as required.

We will now complete the proof of Proposition 5.1. Let $r = \operatorname{rank} G$, $d = \dim G$, and $n = \dim_{\mathbb{F}_2} J$, i.e., $j \simeq \mathbb{Z}_2^n$. Since $G_{(\alpha)}$ is a maximal torus (Proposition 3.22), it is easy to see that $|z_{(\alpha)}| = 2^r$. It follows from Proposition 3.22 that $|j_{(\alpha)}| = 2^{n-1}$. By Lemma 5.5, $2^r/2^{n-1} \leq 4$, or, equivalently,

$$r \leq n + 1. \tag{5.6}$$

From the formula $d = r(2^n - 1)$ (Proposition 3.22) and the formula for the dimension d of a group G of type B_r or D_r,

$$d = r(2r \pm 1)$$

we obtain

$$2r \pm 1 = 2^n - 1. \tag{5.7}$$

Note that $n > 2$, since when $n = 2$ the Weyl J-group $W = \mathrm{SL}_2(J)$ preserves a skew-symmetric bilinear form. Using this fact together with (5.6) and (5.7), we see that there is only one possibility for r and n, namely, $r = 4$, $n = 3$, and the left-hand side of (5.7) must have a minus sign. This means that G is of type D_4. It can easily be verified directly that the group D_4 has no nonclassical Jordan subgroups.

Thus it remains for us to find all nonclassical Jordan subgroups in groups G of exceptional type. First, we will use the formulas in Proposition 3.22 and Lemma 5.2 to find all possible values of p (the period of the Jordan subgroup j) and $n = \dim_{\mathbb{F}_p} J$ for each of the groups G.

LEMMA 5.8. *Suppose j is a nonclassical Jordan subgroup of G, p is its period, and $n = \dim_{\mathbb{F}_p} J$, i.e., $j \simeq \mathbb{Z}_p^n$. Then all possible sets (type G, p, n) are contained in the following list:*

(1) $G = G_2$, $p = 2$, $n = 3$;
(2) $G = F_4$, $p = 3$, $n = 3$;
(3) $G = E_6$, $p = 3$, $n = 3$;
(4) $G = E_8$, $p = 5$, $n = 3$;
(5) $G = E_8$, $p = 2$, $n = 5$.

PROOF. For each simply connected, simple group $\widehat{G}$ we know all primes p such that $\widehat{G}$ contains a subgroup isomorphic to $\mathbb{Z}_p \times \cdots \times \mathbb{Z}_p$ not lying in any torus (see [7]). In view of Lemma 5.2, we see that the required numbers p are exactly the following:

for $G = G_2$, $p = 2$;
for $G = F_4, E_6, E_7$, $p = 2, 3$;
for $G = E_8$, $p = 2, 3, 5$.

In Proposition 3.22 we obtained formulas for the dimension d and rank r of a group G containing a nonclassical Jordan subgroup $j \simeq \mathbb{Z}_p^n$:

$$d = m(p^n - 1); \qquad r = m(p - 1), \quad \text{where } m > 1. \tag{5.9}$$

Knowing the possible values for p and the dimension d and rank r for each exceptional Lie group G, we can use the formulas (5.9) to trivially find all possible values of n and m. The lemma is proved.

We will now prove that for each set (type G, p, n) in Lemma 5.8 there exists a unique nonclassical Jordan subgroup $j \simeq \mathbb{Z}_p^n$ of G.

THEOREM 5.10. *Suppose j is a nonclassical Jordan subgroup of G. Then all nonclassical Jordan subgroups of connected simple algebraic groups, to within conjugacy in G, are given in Table 2. For each nonclassical Jordan subgroup j there is also indicated in Table 2 the factor group z/j, where $z = Z_G(j)$, and the common dimension m of the homogeneous subspaces $\mathfrak{J}_\alpha$ in the grading $\mathfrak{J} = \Sigma^{\boldsymbol{\cdot}}_{\alpha \in J^* \setminus \{0\}} \mathfrak{J}_\alpha$ of the Lie algebra $\mathfrak{J}$ of G.*

TABLE 2. Nonclassical Jordan subgroups

	G	j	W	$Z_G(j)/j$	$\dim_{\mathbb{C}} \mathfrak{J}_\alpha$, $\alpha \in \Omega$
1	G_2	$\mathbb{Z}_2^3$	$SL_3(F_2)$	1	2
2	F_4	$\mathbb{Z}_3^3$	$SL_3(F_3)$	1	2
3	E_6	$\mathbb{Z}_3^3$	$SL_3(F_3)$	$\mathbb{Z}_3^3$	3
4	E_8	$\mathbb{Z}_3^3$	$SL_3(F_5)$	1	2
5	E_8	$\mathbb{Z}_2^5$	$SL_5(F_2)$	$\mathbb{Z}_2^{10}$	8

PROOF. Consider first the cases (1), (2), (4) in Lemma 5.8. Suppose j is a nonclassical Jordan subgroup of $G = G_2$ ($G = F_4$, $G = E_8$, respectively), its period is $p=2$ ($p=3$, $p=5$, respectively), and $j\simeq\mathbb{Z}_p^n$, where $n=3$ ($n=3$, $n=3$, respectively). By Lemma 5.2, each element $x\in j\backslash\{e\}$ lies in the center of some semisimple subgroup H of G of maximal rank. We may obviously assume that $H=Z_G(x)^0$. Since the group G in the cases under consideration is simply connected, we have $Z_G(x)^0=Z_G(x)$, hence $H=Z_G(x)$. From the description of all such subgroups H and their centers (see [12]), and knowing the period p of x, we find that

$$H=(\mathrm{SL}_p\times\mathrm{SL}_p)/\{(\lambda e,\lambda e)|\lambda^p=1\},$$

where $p=2$ for $G=G_2$ ($p=3$ for $G=F_4$, $p=5$ for $G=E_8$).

It is easy to show that there exists in H a commutative subgroup h, unique to within conjugacy, that does not lie in any torus and that contains the center $c(H)=\{x\}$ of H. In addition, $h\simeq\mathbb{Z}_p^3$ and h coincides with its centralizer, $Z_H(h)=h$. It is clear that if j is a nonclassical Jordan subgroup of G with a given set (type G, n, p), then it coincides with h, $j=h$. It remains to show that h is a Jordan subgroup. Let us determine the group $N_G(h)$. First of all, it is easy to show that $N_H(h)/h=\mathrm{SL}_3(h)_x$, where $N_H(h)=N_G(h)\cap H$ and $\mathrm{SL}_3(h)_x$ is the stationary subgroup of x in the group $\mathrm{SL}_3(h)$. It is also clear that the action of $N_G(h)$ on $h\backslash\{e\}$ by conjugation is transitive. Indeed, all nonidentity elements of h are conjugate in G, and all subgroups $h'\subset H$ not lying in a torus and isomorphic to $\mathbb{Z}_p^3$ are conjugate in H. Thus $N_G(h)/h=\mathrm{SL}_3(h)$. It is easy to see from what has been proved that the group $N_G(h)$ is irreducible in the representation by adjoint operators in the Lie algebra $\mathfrak{J}$ of G. Therefore h is a nonclassical Jordan subgroup. Thus Theorem 5.10 has been proved for cases (1), (2), (4). The remaining cases in Lemma 5.8 are handled by analogous arguments. We will not give all of the details. We will mention only certain embeddings of Jordan subgroups and examine in more detail the most complicated case (5) of Lemma 5.8.

A nonclassical subgroup $j\simeq\mathbb{Z}_3^3\subset G=E_6$ (case (3) of Lemma 5.8) can be obtained as follows. The group E_6 contains a subgroup L of type F_4. Take a nonclassical Jordan subgroup j of $L=F_4$. This subgroup j, viewed as a subgroup of $G=E_6$, is a nonclassical Jordan subgroup of G. Its normalizer $N_G(j)$ in G can be extended, in comparison with $N_L(j)$, only by element of the centralizer of j, i.e., $N_G(j)/Z_G(j)=N_L(j)/j$.

We now consider in more detail the case (5): $G=E_8$, $j=\mathbb{Z}_2^5$. Suppose $x\in j\backslash\{e\}$. It follows from Lemma 5.2 and the classification of centers of semisimple subgroups of maximal rank that the subgroup $H=Z_G(x)$ can only be one of two: $H\simeq\mathrm{Spin}'_{16}$ or $H\simeq(\mathrm{SL}_2\times\widehat{E}_7)/\mathbb{Z}_2$. We will prove that in fact $H\simeq\mathrm{Spin}'_{16}$. Assume $H=(\mathrm{SL}_2\times\widehat{E}_7)/\mathbb{Z}_2$. Since $N_G(j)/Z_G(j)\simeq\mathrm{SL}(J)$ for any nonclassical Jordan subgroup j, all nonidentity elements $x\in j\backslash\{e\}$ are conjugate in G. Now consider the representation $\mathrm{Ad}|_j$ of j by adjoint operators in the Lie algebra $\mathfrak{J}$ of G. Since, by definition, $Z_G(j)^0=\{e\}$, it follows that the multiplicity of the unit representation in $\mathrm{Ad}|_j$ is zero. On the other hand,

$$\begin{aligned}\langle 1\,, \mathrm{Ad}|_j\rangle &= \frac{1}{2^5}\{\dim_{\mathbb{C}}\mathfrak{J} + (2^5-1)\cdot[\dim_{\mathbb{C}}\mathscr{L} - (\dim_{\mathbb{C}}\mathfrak{J} - \dim_{\mathbb{C}}\mathscr{L})]\} \\ &= \frac{1}{2^5}\{248 + (2^5-1)[2(133+3) = 248]\} > 0\,.\end{aligned}$$

We have arrived at a contradiction. Thus $H \neq (\mathrm{SL}_2 \times \widehat{E}_7)/\mathbb{Z}_2$. The same argument for the group $H = \mathrm{Spin}'_{16}$ does not lead to a contradiction. We can now show that the group $H = \mathrm{Spin}'_{16}$ contains a unique subgroup $h \simeq \mathbb{Z}_2^5$ such that $Z_H(h)^0 = \{e\}$ and $N_H(h)/Z_H(h) = \mathrm{SL}_5(h)_x$. It is easy to see that this subgroup h is a nonclassical Jordan subgroup of $G = E_8$. The theorem is proved.

Bibliography

1. A. V. Alekseevskiĭ, *On commutative Jordan subgroups of complex Lie groups*, Funktsional. Anal. i Prilozhen. **8** (1974), no. 4, 1–4; English transl. in Functional Anal. Appl. **8** (1974), no. 4.

2. A. E. Zaleskkiĭ and V. N. Serezhkin, *Linear groups generated by transvections*, Izv. Akad. Nauk SSSR Ser. Math. **40** (1976), 26–49; English transl. in Math. USSR–Izv. **10** (1976).

3. V. G. Kats, *Graded Lie algebras and symmetric spaces*, Funktsional. Anal. i Prilozhen. **2** (1968), no. 2, 93–94. (Russian)

4. C. Curtis and I. Reiner, *Representation Theory of Finite Groups and Associative Algebras*, Interscience, New York, 1962.

5. J.-P. Serre, *Lie Groups and Lie Algebras*, Benjamin, New York, 1965.

6. D. A. Suprunenko, *Soluble and Nilpotent Linear Groups*, Minsk, 1958; English transl., Transl. Math. Monographs, vol. 9, Amer. Math. Soc., Providence, RI, 1963.

7. A. Borel, *Sous-groupes commutatifs et torsion des groupes de Lie compact connexes*, Tôhoku Math. J. **13** (1961), no. 2, 216–240. (Russian)

8. J. McLaughlin, *Some groups generated by transvections*, Arch. Math. (Basel) **18** (1967), no. 4, 364–368.

9. ——, *Some subgroups of* $\mathrm{SL}_n(F_2)$, Illinois J. Math. **13** (1969), 108–115.

10. H. Pollatsek, *Groups generated by transvections over the field of two elements*, J. Algebra **16** (1970), 561–574.

11. I. Popovici, *Graduations spéciales simple*, Bull. Soc. Roy. Sci. Liège **39** (1970), nos. 5–6, 218–228.

12. J. de Siebental, *Sur certain modules dans une algèbre de Lie semi-simple*, Comment. Math. Helv. **44** (1969), 1–44.

13. P. E. Smith, *A simple subgroup of M ? and* $E_8(3)$, Bull. London Math. Soc. **8** (1976), no. 2, 161–165.

14. J. Tits, *Sur les constantes de structure et le théorème d'existence des algèbres de Lie semi-simples*, Inst. Hautes Études Sci. Publ. Math. (1966), no. 31, 21–58.

15. J. G. Thompson, *A conjugacy theorem for* E_8, J. Algebra **38** (1976), 525–530.

16. D. Wales, *Finite linear groups of degree seven*. II, Pacific J. Math. **34** (1970), 207–235.

17. C. Jordan, *Oeuvres*, vol. 2, Paris, 1961.

18. Wim H. Hesselink, *Special and pure gradings of Lie algebras*, Math. Z. **179** (1982), no. 1, 135–149.

19. A. I. Kostrikin, I. A. Kostrikin, and V. A. Ufnarovskiĭ, *Orthogonal decompositions of simple Lie algebras*, Dokl. Akad. Nauk SSSR **260** (1981), 526–530; English transl. in Soviet Math. Dokl. **24** (1981).

20. ——, *Multiplicative decompositions of simple Lie algebras*, Dokl. Akad. Nauk SSSR **262** (1982), 29–33; English transl. in Soviet Math. Dokl. **25** (1982).

21. ——, *Decompositions of classical Lie algebras*, Trudy Mat. Inst. Steklov. **166** (1984), 107–122; English transl. in Proc. Steklov Inst. Math. **166** (1986).

22. V. P. Burichenko, *Transitive orthogonal decompositions of simple complex Lie algebras of types F_4 and E_6*, Vestnik Moskov. Univ. Ser. 1 Mat. Mekh. **4** (1988), 78–80; English transl. in Moscow Univ. Math. Bull. **43** (1988).

23. A. V. Alekseevskiĭ, *Maximal solvable subgroups of Lie groups*, Funktsional Anal. i Prilozhen. **14** (1980), no. 2, 44–45; English transl. in Functional Anal. Appl. **14** (1980), no. 2.

24. ([6]) ——, a) *Maximal finite subgroups of Lie groups*, Funktsional. Anal. i Prilozhen. **9** (1975), no. 3, 79–80; English transl. in Functional Anal. Appl. **9** (1975), no. 3.

b) *The structure of maximal finite primitive subgroups of Lie groups*, Uspekhi Mat. Nauk **30** (1975), no. 5 (185), 197–198. (Russian)

Translated by G. A. KANDALL

([6]) Editor's note. The bibliography of the Russian original does not include an item 24. The author may have intended one or both of the papers cited here.

Amer. Math. Soc. Transl.
(2) Vol. **151**, 1992

On a Noncommutative Analogue of the Category of Coherent Sheaves on a Projective Scheme

UDC 512.732

A. B. VEREVKIN

Introduction and Main Results

In his classical paper [1], J.-P. Serre introduced and studied the cohomology of (quasi) coherent sheaves on projective algebraic varieties. In particular, he was able to characterize those sheaves in a purely algebraic way: the category $\mathrm{Coh}(X)$ on the projectivized spectrum $\operatorname{Proj} A$ of a commutative graded k-algebra A is equivalent to the category of graded modules over A modulo finite-dimensional modules (see below for details).

In this paper we study, following a suggestion of Yu. I. Manin, similar categories constructed from not necessarily commutative rings and we also prove several results analogous to the main theorems of Serre.

Thus we construct a fragment of "noncommutative algebraic geometry" bypassing the geometry in the narrow meaning of the word. Nowadays, after several and not very successful attempts to imitate the definition of a scheme with the aid of noncommutative localization, an old idea of Grothendieck is beginning to prevail. According to it, it is not necessary to have a "noncommutative space": a good category of sheaves on the space should do as well. As additional evidence in support of this approach, one could mention the results of A. Connes [2], who laid the foundation of noncommutative differential geometry, the developments in supergeometry [3], as well as the recent results [4], [5] describing (derived) categories of sheaves, which show that even usual varieties can naturally lead to modules over quite noncommutative rings.

Before we discuss in detail what we have done in the noncommutative case we would like to recall the results of Serre.

1991 *Mathematics Subject Classification.* Primary 14F05.

Serre considered a commutative graded k-algebra A generated by its graded component A_1, where A_1 was assumed to be a finite-dimensional k-vector space.

With this algebra A there are associated the following structures:

(a) the category $\mathscr{M}^0$ of "almost" finitely generated graded A-modules, more precisely,

$$M = \bigoplus_{i\in\mathbb{Z}} M_i \in \operatorname{Ob}\mathscr{M}^0$$

if and only if there exists an integer n such that $M_{\geq n} := \bigoplus_{i\geq n} M_i$ is a finitely generated graded A-module;

(b) the dense (i.e., full and closed under kernels, cokernels and extensions) subcategory $\mathscr{C}$ of $\mathscr{M}^0$ consisting of modules with only finitely many nonzero components of positive degree;

(c) the projective algebraic scheme $X := \operatorname{Proj} A$;

(d) the category $\operatorname{Coh}(X)$ of coherent sheaves on X.

The following theorems were proved by Serre in [1].

THEOREM 1. $\operatorname{Coh}(X) \simeq \mathscr{M}^0/\mathscr{C}$ *where the quotient category is defined in the sense of Grothendieck* [6]. *The equivalence is given by the functor* $\mathscr{F} \mapsto \Gamma(\mathscr{F}) := \bigoplus_{i\in\mathbb{Z}} H^0(X, \mathscr{F}(i))$.

THEOREM 2. *For any sheaf* $\mathscr{F} \in \operatorname{Ob}(\operatorname{Coh}(X))$, *there are isomorphisms*

$$H^i(X, \mathscr{F}) \cong \operatorname{Ext}^i(X; O_X, \mathscr{F}) \cong \operatorname{Ext}^i_{\mathscr{M}^0/C}(A, \Gamma(\mathscr{F})).$$

THEOREM 3 (On the cohomology of Serre sheaves).

$$H^{\cdot}(X, Q_X(j)) = H^0(X, O_X(j)) \cong A_j \quad \textit{for all } j \gg 0.$$

THEOREM 4. *For any sheaf* $\mathscr{F} \in \operatorname{Ob}(\operatorname{Coh}(X))$ *there exists a* k *such that for any integer* $i \geq k$ *we have that*

$$H^{\cdot}(X, \mathscr{F}(i)) = H^0(X, \mathscr{F}(i)).$$

THEOREM 5. $\operatorname{Coh}(X) \simeq \operatorname{Coh}(X^{(d)})$ *for any positive integer* d, *where*

$$X^{(d)} := \operatorname{Proj}(A^{(d)}) := \operatorname{Proj}\left(\bigoplus_{i\geq 0} A_{d\cdot i}\right)$$

(*this follows from the fact that the embedding* $A^{(d)} \hookrightarrow A$ *induces an isomorphism* $\operatorname{Proj} A \cong \operatorname{Proj} A^{(d)}$).

We shall now consider the noncommutative case. Let $A = \bigoplus_{i\geq 0} A_i$ be a not necessarily commutative graded left Noetherian ring with identity, $\mathscr{M}$ the category of graded left A-modules containing both the full subcategory $\mathscr{M}^0$ of "almost" finitely generated modules and the dense subcategory $\mathscr{C}$ consisting of A-modules with only finitely many nonzero components of positive degree. We want to introduce yet another dense subcategory $\overline{\mathscr{C}}$ of $\mathscr{M}$.

It consists of colimits of functors from directed categories to $\mathscr{C}$ and it is closed under inductive limits.

To categories $\mathscr{C}$ and $\overline{\mathscr{C}}$ there correspond two sets of arrows: Σ, consisting of $\mathscr{C}$-isomorphisms (whose kernels and cokernels belong to $\mathrm{Ob}\,\mathscr{C}$), and $'\Sigma$, consisting of $\overline{\mathscr{C}}$-isomorphisms. Noncommutative geometry proper begins when we declare the categories $\mathscr{M}^0\Sigma^{-1}$ and $\mathscr{M}'\Sigma^{-1}$ the category Coh of coherent and, respectively, the category Qco of quasicoherent sheaves. The groups $\mathrm{Ext}^i_{\mathrm{Qco}}(A, F)$ will be called the cohomology of a sheaf $F \in \mathrm{Ob}\,\mathrm{Coh}$ (resp. $\mathrm{Ob}\,\mathrm{Qco}$).

Thus Theorems 1 and 2 are now definitions and our goal is to obtain analogues of the remaining theorems from these definitions.

The first important facts are that the new objects of study are well defined, the natural functor $\mathrm{Coh} \mapsto \mathrm{Qco}$ is a full embedding and Qco has enough injectives, the last property being quite useful for computation of cohomology by the formula

$$H^i(F) \cong \varinjlim_{\mathbb{Z}} \mathrm{Ext}^i_{\mathscr{M}}(A_{\geq n}, F).$$

Notice that, except for some trivial cases, the categories Coh and Qco have no projectives.

It is convenient to replace cohomology groups with cohomology modules $\underline{H}^j(F) := \bigoplus_{i\in\mathbb{Z}} H^j(F(i))$, which are objects of $\mathscr{M}$.

A partial analogue of Theorems 3 and 4 can be stated as follows.

THEOREM (A-3,4). (a) *If $F \in \mathrm{Ob}\,\mathscr{M}$ then the natural morphism*

$$\underline{\mathrm{Hom}}_{\mathscr{M}}(A, F) \cong F \to \underline{H}^0(F)$$

in $\mathscr{M}$ belongs to the class $'\Sigma$.

(b) *If $F \in \mathrm{Ob}\,\mathscr{M}$ and i is a positive integer then*

$$\underline{H}^i(F) := \underline{\mathrm{Ext}}^i_{\mathrm{Qco}}(A, F) \in \mathrm{Ob}\,\overline{C}.$$

For a complete analogy with Theorems 3 and 4 we should have proved that $\underline{H}^j(F) \in \mathrm{Ob}\,\mathscr{M}^0$, where $F \in \mathrm{Ob}\,\mathscr{M}^0$. However, Theorem (A-3,4) has a "geometric meaning" since a $\overline{\mathscr{C}}$-isomorphism of modules implies the isomorphism of localizations of these modules with respect to any set of denominators admissible in the given situation and containing elements of positive degree.

To obtain stronger results about cohomology one can impose a certain finiteness condition on the ring A:

THEOREM (A-6). *Let $\mathscr{M}$ be the category of left A-modules and suppose that* l. gl. $\dim A = n > 0$. *Then for any integer $i \geq n$ and quasicoherent sheaf F we have that $H^i(F) = 0$.*

The following analogue of Theorem 5 about the Veronese equivalence can be proved in a rather general form:

THEOREM (A-5). *Let $A = A_0\langle A_1\rangle$ be a graded left Noetherian ring. Then the categories of coherent and quasicoherent sheaves, obtained by localizing the categories of left $A^{(d)}$- and A-modules, are equivalent for any natural d.*

This "Veronese equivalence" may be useful since it reduces the fragment of noncommutative geometry under consideration to the study of algebras defined by quadratic relations. Indeed, as was shown in [7], the algebra $A^{(d)}$ has this property for sufficiently large d. The duality theory for those also shows that there is a second, and rather interesting, possibility for a definition of $\mathrm{Coh}(\text{“}X\text{”})$: to mod out projective A-modules rather than finite-dimensional ones. This category, as was noted by Yu. I. Manin and R.-O. Buchweitz, deserves to be studied even for the usual supervarieties (over smooth varieties it is trivial). For details see [5].

§2. Categories Coh and Qco are well defined

Let B be an abelian category and $S \subset \mathrm{Mor}\, B$. Then S admits calculus of left fractions ([8]) if the following conditions are satisfied:

(a) S is closed under composition and contains all identity morphisms,

(b) for any $f \in B(X, Y)$ and $s \in B(X, Z)\cap S$ there exist $t \in B(Y, W)\cap S$ and $g \in B(Z, W)$ such that $t \circ f = g \circ s$.

(c) If for a pair of morphisms $f, g \in B(X, Y)$ there exists a right equalizer $s \in B(Z, X)\cap S$ (i.e., $f\circ s = g\circ s$) then there also exists a left equalizer $t \in B(Y, W) \cap S$.

By inverting the arrows we get similar axioms for calculus of right fractions. We shall later see that both Σ in $\mathcal{M}$ and $\Sigma \cap \mathcal{M}^0$ in $\mathcal{M}^0$ admit calculus of left, as well as right, fractions. Hence there exist universal functors from those categories into abelian categories which invert the aforementioned classes of arrows.

LEMMA 2.0. *If $W \in \mathrm{Ob}\,\mathcal{M}$ then there exists $X \subset W$ such that $\overline{C}(X) := \{x \in X | A\cdot x \in \mathrm{Ob}\, C\}$ and $W/X \in \mathrm{Ob}\,\overline{C}$.*

PROOF. For such a submodule we can take, by Zorn's lemma, a maximal element of

$$\{Y \hookrightarrow W / Y \cap \overline{C}(W) = 0\}.$$

LEMMA 2.1. *Let $f \in \mathrm{Mor}\,\mathcal{M}$ and $\mathrm{Ker}\, f \in \mathrm{Ob}\,\overline{C}$. Then there exists $X \hookrightarrow \mathrm{dom}(f)$ such that $f|_X$ is a monomorphism in $\mathcal{M}$.*

Moreover $f \in \Sigma$ if and only if there exists an integer n such that f is an isomorphism in all degrees greater than 2.

PROOF. The module from Lemma 2.0 can be taken for X. The second assertion is obvious.

LEMMA 2.2. $\mathrm{Ob}\overline{\mathscr{C}} = \{Z \in \mathrm{Ob}\mathscr{M} \mid$ *any morphism from a cyclic free module to* Z *factors through an object of* $\mathscr{C}\}$.

PROOF. If $Z \in \mathrm{Ob}\overline{\mathscr{C}}$ then there exist a directed category I and F in $\mathrm{Ob}\mathscr{C}^I$ such that $Z = \mathrm{colim}\, F$. For each $i \in \mathrm{Ob}\, I$ we set $\mathscr{C}_i := F(i)$ and let $\gamma_i \in \mathscr{M}(\mathscr{C}_i, Z)$ denote the canonical morphism. Then $Z \cong \varinjlim \mathrm{Im}\,\gamma_i \cong U_I \mathrm{Im}\,\gamma_i$. For $\alpha \in \mathscr{M}(A(k), Z)$ we set $\alpha(1(k)) = z \in Z$. We can then find $j \in \mathrm{Ob}\, I$ such that $z \in \mathrm{Im}\,\gamma_i$. For $c \in C_j \cap \gamma_j^{-1}(Z)$ we have that $\beta\colon A(k) \to C_j\colon a(k) \to a\cdot c$. It is clear that $\beta \in \mathscr{M}(A(k), C_j)$ and we have the identities

$$d(a(k)) = a\cdot d(1(k)) = a\cdot Z = a\cdot\gamma_j(c) = a\cdot\gamma_j(\beta(1(k))) = \gamma_j \circ \beta(a(k)).$$

Therefore $d = \gamma_j \circ \beta$ and we have the inclusion $\mathrm{Ob}\,\overline{C} \subset \{\cdots\}$.

Let $Z \in \{\cdots\}$ and $\{z_i\}_{i\in I}$ be an ordered system of homogeneous generators of Z. Consider the morphisms

$$d_i\colon A(k) \to Z\colon d_i(1(K_i)) = Z_i,\quad k_i = \deg Z_i.$$

Then $Z = \bigcup_I \mathrm{Im}\,\alpha_i$ and, by hypothesis, $\mathrm{Im}\,\alpha_i \in \mathrm{Ob}\, C$ for each $i \in \mathrm{Ob}\, I$. We set $C_i := \mathrm{Im}\,\alpha_i$ and let $\gamma_i\colon C_i \to Z$ denote the embedding. For any natural k we also set

$$C^{(i_1,\dots,i_k)} := \bigoplus_{j=1}^{k} C_{i_j} \in \mathrm{Ob}\, C, \qquad \gamma^{(i_1,\dots,i_k)} := \bigoplus_{j=1}^{k} \gamma_{i_j}.$$

Notice that unordered tuples of elements of an ordered set form a directed category L.

Consider the functor $F \in \mathrm{Ob}\, C^L$; $F(l) = C^{(l)}$. By construction, $Z \simeq \mathrm{colim}\, F$ and the assertion of the lemma follows.

COROLLARY 2.3. $\mathrm{Ob}\,\overline{C} = \{Z \in \mathrm{Ob}\mathscr{M} \mid$ *any cyclic submodule of* Z *belongs to* $C\}$.

LEMMA-COROLLARY 2.4. $\mathrm{Ob}\,\overline{C} = \{Z \in \mathrm{Ob}\mathscr{M} \mid$ *for any* $M \in \mathrm{Ob}\mathscr{M}^0$ *and* $d \in \mathscr{M}(M, Z)$, d *factors through an object of* $C\}$.

The proof follows from 2.3.

COROLLARY 2.5. $\mathrm{Ob}\,\overline{C} = \{Z \in \mathrm{Ob}\mathscr{M} \mid$ *any "almost" finitely generated submodule of* Z *belongs to* $C\}$.

LEMMA 2.6. (a) $\overline{C}$ *is a dense subcategory of* $\mathscr{M}$,

(b) $\overline{C}$ *is closed under inductive limits.*

PROOF. It is clear that $\overline{C}$ is closed under subobjects in the category $\mathscr{M}$. By Lemma 2.2, we have that $\overline{C}$ is closed under images in the category $\mathscr{M}$.

Consider an extension $0 \to Z_1 \to X \xrightarrow{d} Z_2 \to 0$ in $\mathcal{M}$, where $Z_1, Z_2 \in \operatorname{Ob}\overline{C}$. Let M be a cyclic submodule of X in $\mathcal{M}$. Then the sequence

$$0 \to \operatorname{Ker}(\alpha|_M) \to M \xrightarrow{\alpha|M} \alpha(M) \to 0$$

is exact in $\mathcal{M}$. By Corollary 2.3, $\alpha(M) \in \operatorname{Ob} C$. Since A is Noetherian, we have that $\operatorname{Ker}(\alpha|_M)$ is finitely generated and, by Corollary 2.5, it belongs to $\operatorname{Ob} C$. Therefore $M \in \operatorname{Ob} C$ and $X \in \operatorname{Ob}\overline{C}$.

Let I be a directed category, $F \in \operatorname{Ob}\overline{C}_I$ and $X = \operatorname{colim} F$. Then any morphism from a free module to X can be factored through some $F(i)$, where $i \in \operatorname{Ob} I$. Therefore $X \in \operatorname{Ob}\overline{C}$ also factors through an object of C.

Recall now that ${}'\Sigma$ was defined as the class of $\overline{C}$-isomorphisms in the category $\mathcal{M}$.

THEOREM 2.7. *${}'\Sigma$ in $\mathcal{M}$ admits calculus of left and right fractions and therefore there exists a universal ${}'\Sigma$-inverting functor*

$$P_{'\Sigma}\colon \mathcal{M} \mapsto \mathcal{M}'\Sigma^{-1} \simeq {}'\Sigma^{-1}\mathcal{M}$$

into an abelian category.

PROOF. The result follows from the fact that subcategory $\overline{C}$ is dense in $\mathcal{M}$ ([8]).

We shall now set new notations. Let B be an abelian category and S a set of arrows in B admitting calculus of left fractions. We then set

$$\operatorname{Ob} S^{-1}B := \operatorname{Ob} B$$

$$S^{-1}B(X, Y) := \{(f, \sigma) := (X \xrightarrow{f} y' \xleftarrow{\sigma} y),\ \underline{\sigma} \in S\}$$

where $(f, \underline{\sigma}) = (g, \underline{\tau})$, if there exist $W \in \operatorname{Ob} B$, $\underline{\alpha} \in B(\operatorname{range}(\underline{\sigma}), W)$ and $\beta \in B(\operatorname{range}(\underline{\tau}), W)$ such that $\alpha \circ f = \beta \circ g$ and $\underline{\alpha \circ \sigma} = \underline{\beta \circ \tau} \in S$. Composition in $S^{-1}B$ is then defined via axiom (b) similar to the product in the left localization of a ring with respect to an Ore set.

Let S consist of $\mathcal{D}$-isomorphisms, where $\mathcal{D}$ is a dense subcategory of B. Then S admits calculus of left and right fractions and we have the following result.

LEMMA 2.8. (a) $(\underline{\sigma}, f) = 0$ *in* BS^{-1} *if and only if* $\operatorname{Im}(f) \in \operatorname{Ob}\mathcal{D}$,
(b) $(f, \underline{\sigma}) = 0$ *in* $S^{-1}B$ *if and only if* $\operatorname{Im}(f) \in \operatorname{Ob}\mathcal{D}$.

PROOF. (a) $(\sigma, f) = (1, 0)$ and we can find $W \in \operatorname{Ob}$, $\mu \in B(W, \operatorname{dom}(f))$, and $\tau \in B(W, \operatorname{ran}(\sigma))$ such that $\sigma \circ \mu = \tau \in S$ and $f \circ \mu = 0$. It follows from the first equation that the sequence $\pi(\operatorname{Ker}(\sigma)) \mapsto \operatorname{Coker}(\mu) \to \operatorname{Coker}(\tau)$, where $\pi \in B(\operatorname{dom}(\sigma), \operatorname{Coker}(\mu))$ is the canonical projection, is exact and therefore $\operatorname{Coker}(\mu) \in \operatorname{Ob}\mathcal{D}$. It follows from the second equation that $\operatorname{Coker}(\mu) \to \operatorname{Im}(f)$ is an epimorphism and therefore $\operatorname{Im}(f) \in \operatorname{Ob}\mathcal{D}$.

Conversely, if $\mathrm{Im}(f) \in \mathrm{Ob}\mathscr{D}$ then $\pi\colon Y \to \mathrm{Coker}(f) \in S$ and $(\sigma, f) = (\pi, 1) \circ (\sigma, 0) = 0$.

The proof of part (b) is similar to that of (a).

Returning now to the categories Coh and Qco we see that the inclusions $\Sigma \subset {}'\Sigma$ and $\mathscr{M}^0 \subset \mathscr{M}$ yield two functors

$$F\colon \mathscr{M}^0\Sigma^{-1} \mapsto \mathscr{M}\Sigma^{-1}, \qquad G\colon \mathscr{M}\Sigma^{-1} \mapsto \mathscr{M}'\Sigma^{-1}.$$

LEMMA 2.9. (a) *F is full and faithful,*

(b) *G is exact,*

(c) *$G \circ F$ is full and faithful.*

PROOF. Parts (a) and (b) are obvious; part (c) follows from Lemma 2.10 below.

It follows from Lemma 2.9 that ([1]) is a full embedding.

LEMMA 2.10. *Let $y \in \mathrm{Ob}\mathscr{M}$ and $x \in \mathrm{Ob}\mathscr{M}^0$. Then*

$$\mathrm{Hom}_{\mathscr{M}\Sigma^{-1}}(x, y) \cong \mathrm{Hom}_{\mathscr{M}'\Sigma^{-1}}(x, y).$$

PROOF. Consider the homomorphism

$$g\colon \mathrm{Hom}_{\mathscr{M}\Sigma^{-1}}(x, y) \to \mathrm{Hom}_{\mathscr{M}'\Sigma^{-1}}(x, y),$$

$$\mathrm{Ker}(g) = \{(\sigma, f) \in \mathscr{M}\Sigma^{-1}(x, y) : \mathrm{Im}(f) \in \mathrm{Ob}\,\overline{C}\}.$$

Since $x \in \mathrm{Ob}\mathscr{M}^0$ and $\sigma \in \Sigma$, we have that $\sigma^{-1}(x) \in \mathrm{Ob}\mathscr{M}^0$ and therefore $\mathrm{Im}(f) \in \mathrm{Ob}\mathscr{M}^0$. It is clear that $\mathrm{Ob}\mathscr{M}^0 \cap \mathrm{Ob}\,\overline{C} = \mathrm{Ob}\,C$ whence $\mathrm{Ker}(g) = 0$.

Let (σ, f), where $\sigma \in {}'\Sigma$, be a representative of a morphism from the right-hand side of

$$\mathrm{Hom}_{\mathscr{M}\Sigma^{-1}}(x, y) \cong \mathrm{Hom}_{\mathscr{M}'\Sigma^{-1}}(x, y).$$

We may assume that σ is an epimorphism and x is finitely generated. Then we can find a finitely generated submodule M of $\mathrm{dom}(\sigma)$ such that $\sigma|_M$ is an epimorphism. Since $(\sigma, f) = (\sigma|_M, f|_M)$ we have that g is an epimorphism.

LEMMA 2.11. *If $Y \in \mathrm{Ob}\mathscr{M}$ and $X \in \mathrm{Ob}\mathscr{M}^0$ then for each i we have the isomorphisms*

$$\mathrm{Ext}^i_{\mathscr{M}\Sigma^{-1}}(X, Y) \cong \mathrm{Ext}^i_{\mathrm{Qco}}(X, Y).$$

PROOF. We shall define two covariant functors

$$h^X_\Sigma\colon \mathscr{M}\Sigma^{-1} \mapsto \mathrm{Ab}, \qquad h^X_{'\Sigma}\colon \mathrm{Qco} \mapsto \mathrm{Ab}.$$

On objects they are defined as follows:

$$h^X_\Sigma(Y) := \mathrm{Hom}_{\mathscr{M}\Sigma^{-1}}(X, Y), \qquad h^X_{'\Sigma}(Y) := \mathrm{Hom}_{\mathrm{Qco}}(X, Y).$$

On arrows they are defined in the natural way. By Lemma 2.10, we have that $h^X_\Sigma = h^X_{'\Sigma} \circ G$ and $Rh^X_\Sigma = Rh^X_{'\Sigma} \circ G$, since G is exact, and by 2.1 and 2.8, the

([1]) Editors's note. There is an omission at this point in the Russian original; the author may have intended $\mathscr{M}^0\Sigma^{-1} \mapsto \mathscr{M}'\Sigma^{-1}$.

universal, with respect to inversion of arrows, functors P_Σ and P'_Σ preserve injectivity.

If $\mathcal{Y} \to I^\circ$ be an injective resolution in $\mathcal{M}$ then it is also an injective resolution in categories $\mathcal{M}\Sigma^{-1}$ and Qco and therefore it is also an acyclic, with respect to the functor h'^X_Σ, resolution in the category Qco.

Therefore, for any i, we have the isomorphisms

$$\mathrm{Ext}^i_{\mathcal{M}\Sigma^{-1}}(X, Y) \cong H^i(h^X_\Sigma(I^\circ)) \cong H^i(h'^X_\Sigma(Y)) \cong \mathrm{Ext}^i_{\mathrm{Qco}}(X, Y). \quad \square$$

Associated with $Y \in \mathrm{Ob}\mathcal{M}$ there are the submodules

$$y^{(i)} := \{y \in Y | A_{\geq i} \circ y = 0\}, \qquad i \in \mathbb{N}.$$

Let $Y^{(\infty)} = \bigcup_{i\in\mathbb{N}} y^{(i)}$. By Corollary 2.3, for all $i \in \mathbb{N}$, we have that $y^{(i)} \in \mathrm{Ob}\,\overline{C}$ and therefore $Y^{(\infty)} \in \mathrm{Ob}\,\overline{C}$.

If $Y \in \mathrm{Ob}\mathcal{M}^0$ then for any $i \in \mathbb{N} \cup \{\infty\}$ we have that $Y^{(i)} \in \mathrm{Ob}\, C$.

LEMMA 2.12. *If $Y \in \mathrm{Ob}\mathcal{M}$ then:*

(a) *for any $i \in \mathbb{N} \cup \{\infty\}$ we have that $(Y/Y^{(\infty)})^{(i)} = 0$;*

(b) *for any $C' \in \mathrm{Ob}\,\overline{C}$ we have that* $\mathrm{Hom}_{\mathcal{M}}(C', Y) \cong \mathrm{Hom}_{\mathcal{M}}(C', Y^{(\infty)})$.

PROOF. Part (a) follows from the fact that A is Noetherian and that $Y^{(\infty)} \in \mathrm{Ob}\,\overline{C}$;

(b) It is easy to see that if $Y^{(\infty)} = 0$ then $\mathrm{Hom}_{\mathcal{M}}(C', Y) = 0$. If $Y^{(\infty)} \neq 0$ then consider the sequence

$$0 \to Y^{(\infty)} \to Y \to Y/Y^{(\infty)} \to 0$$

which is exact in $\mathcal{M}$. It follows that the sequence

$$0 \to \mathrm{Hom}_{\mathcal{M}}(C', Y^{(\infty)}) \to \mathrm{Hom}_{\mathcal{M}}(C', Y) \to \mathrm{Hom}_{\mathcal{M}}(C', Y/Y^{(\infty)})$$

is also exact. By part (a), the last group is zero and therefore the first two are isomorphic.

We shall now define inner Hom's and Ext's in $\mathcal{M}$:

$$\underline{\mathrm{Hom}}_{\mathcal{M}}(X, Y) := \bigoplus_{i\in\mathbb{Z}} \mathrm{Hom}_{\mathcal{M}}(X, Y(i)),$$

$$\mathrm{Ext}^k_{\mathcal{M}}(X, Y) := \bigoplus_{i\in\mathbb{Z}} \mathrm{Ext}^k_{\mathcal{M}}(X, Y(i)).$$

If $X \in \mathrm{Ob}\mathcal{M}$ has a structure of graded A-bimodule then the graded groups defined above become objects of $\mathcal{M}$. Indeed, for any $a \in A$ and $f \in \underline{\mathrm{Hom}}_{\mathcal{M}}(X, Y)$ define $(a \circ f) \in \underline{\mathrm{Hom}}_{\mathcal{M}}(X, Y)$ by the formula $(a \circ f)(x) := f(x \circ a)$.

It is not difficult to verify that $\underline{\mathrm{Hom}}_{\mathcal{M}}(X, Y) \in \mathrm{Ob}\mathcal{M}$.

Using injective resolutions one can show that inner Ext's belong to the same categories as inner Hom's.

LEMMA 2.13. *Suppose that $X \in \mathrm{Ob}\,\overline{C}$ is a graded A-bimodule. Then:*

(a) *if $X = \bigoplus_{i\geq d} X_i$ and $Y \in \mathrm{Ob}\mathcal{M}^0$ then* $\underline{\mathrm{Hom}}_{\mathcal{M}}(X, Y) \in \mathrm{Ob}\, C$.

(b) *if* X *is finitely generated in* $\mathscr{M}$ *and* $Y \in \operatorname{Ob}\mathscr{M}$ *then* $\operatorname{Ext}^i_{\mathscr{M}}(X, Y) \in \operatorname{Ob}\overline{C}$ *for each nonnegative integer* i.

PROOF. In both cases $\underline{\operatorname{Hom}}_{\mathscr{M}}(X, Y) \cong \underline{\operatorname{Hom}}_{\mathscr{M}}(X, Y^{(\infty)})$.

(a) We have that $Y^{(\infty)} \in \operatorname{Ob} C$. Therefore if $Y^{(\infty)}_{\geq n} = 0$ then, as is easily seen, $\underline{\operatorname{Hom}}_{\mathscr{M}}(X, Y)_{\geq n-d} = 0$.

(b) In this case the image of any morphism is finitely generated and therefore belongs to $\operatorname{Ob} C$. This means that any morphism is an A_+-torsion element and therefore $\underline{\operatorname{Hom}}_{\mathscr{M}}(X, Y) \in \operatorname{Ob}\overline{C}$. Let $Y \to I^{\cdot}$ be an injective resolution in $\mathscr{M}$. Then

$$\operatorname{Ext}^i_{\mathscr{M}}(X, Y) = H^i\underline{\operatorname{Hom}}_{\mathscr{M}}(X, I^{\cdot})) \in \operatorname{Ob}\overline{C},$$

and each term of the complex belongs to $\operatorname{Ob}\overline{C}$. □

§3. The structure of cohomology

By Lemmas 2.10 and 2.11, cohomological questions reduce to questions about Ext's in the category $\mathscr{M}\Sigma^{-1}$.

We shall now introduce the following definitions:

(1) $\mathbb{Z}$ will be the directed category of integers,

(2) $\mathrm{Ab}^{\mathbb{Z}}$ will denote the category of covariant functors from $\mathbb{Z}$ to Ab,

(3) $\mathscr{M}_{\mathbb{Z}}$ will denote the category of contravariant functors from $\mathbb{Z}$ to $\mathscr{M}$;

(4) for each $M \in \operatorname{Ob}\mathscr{M}$ we define $F_M \in \operatorname{Ob}\mathscr{M}_{\mathbb{Z}}$ by the formulas

$$F_M(i) := M_{\geq i}, \quad F_M(i \to j) := (M_{\geq i} \hookleftarrow M_{\geq j});$$

(5) $\mathbb{F}\colon \mathscr{M} \mapsto \mathscr{M}_{\mathbb{Z}}$ will denote the contravariant functor with $\mathbb{F}(M) := F_M$ and $\mathbb{F}(M \overset{g}{\mapsto} N) :=$ the natural transformation $F_M \to F_N$ of functors defined by the morphism g;

(6) for each $Y \in \operatorname{Ob}\mathscr{M}$ we define the contravariant functor

$$h_Y\colon \mathscr{M} \mapsto \mathrm{Ab} : h_Y(X) = \operatorname{Hom}_{\mathscr{M}}(X, Y),$$

$$h_Y(X \xrightarrow{f} Z) = (\operatorname{Hom}_{\mathscr{M}}(Z, Y) \xrightarrow{f^*} \operatorname{Hom}_{\mathscr{M}}(X, Y)).$$

It is not difficult to see that it can be extended componentwise to a contravariant functor $h_Y\colon \mathscr{M}_{\mathbb{Z}} \mapsto \mathrm{Ab}^{\mathbb{Z}}$.

THEOREM 3.1. *For any* $X, Y \in \operatorname{Ob}\mathscr{M}$ *there are isomorphisms*

$$\operatorname{Hom}_{\mathscr{M}\Sigma^{-1}}(X, Y) \cong \varinjlim (h_Y \circ \mathbb{F})(X) \cong \varinjlim h^{\mathbb{F}(X)}(Y).$$

PROOF. The second isomorphism is obvious. To prove the existence of the first it suffices to remark that the functor $F_X^{\mathrm{op}}\colon \mathbb{Z}^{\mathrm{op}} \mapsto (\Sigma/X)^{\mathrm{op}}$ is cofinal and therefore yields an isomorphism ([9], I.8.4.; [10], A7).

LEMMA 3.2. *The following assertions are true in* Ab:

(a) *the inductive limit of exact sequences is exact,*

(b) $\varinjlim$ *commutes with* Ker *and* Im,

(c) *on complexes,* $\varinjlim$ *commutes with cohomology.*

Proof. Assertion (a) is a consequence of axiom AB5 which holds in Ab ([10], A). Part (b) follows from (a).

Theorem 3.3. *For any* $X, Y \in \mathrm{Ob}\,\mathcal{M}$ *and* $i \in \mathbb{Z}$ *there are isomorphisms*

$$\mathrm{Ext}^i_{\mathcal{M}\Sigma^{-1}}(X, Y) \cong \varinjlim (R^i h_y \circ \mathbb{F})(X) \cong \varinjlim R^i h^{\mathbb{F}(X)}(Y).$$

Proof. Let $Y \to I^\circ$ be an injective resolution in $\mathcal{M}$. It is also an injective resolution in $\mathcal{M}\sigma^{-1}$. By 3.1 and 3.2, we have the isomorphisms

$$\begin{aligned}\mathrm{Ext}^i_{\mathcal{M}\Sigma^{-1}}(X, Y) &\cong H^i(\mathrm{Hom}_{\mathcal{M}\Sigma^{-1}}(X, I^\circ)) \\ &\cong H^i(\varinjlim h^{\mathbb{F}(X)}(I^\circ)) \cong \varinjlim H^i(h^{\mathbb{F}(X)}(I^\circ)) \\ &\cong \varinjlim R^i h^{\mathbb{F}(X)}(Y).\end{aligned}$$

The isomorphism between the second and third groups under consideration is obvious.

In the introduction we stated Theorem (A-3, 4). Its proof is based on the following lemma.

Lemma 3.4. *Suppose that* $X \in \mathrm{Ob}\,\mathcal{M}$ *is finitely generated and is also a graded A-bimodule. Then for each* $Y \in \mathrm{Ob}\,\mathcal{M}$ *the natural morphism* $\underline{\mathrm{Ext}}^i_{\mathcal{M}}(X, Y) \to \underline{\mathrm{Ext}}^i_{\mathcal{M}\Sigma^{-1}}(X, Y)$ *in* $\mathcal{M}$ *belongs to* $'\Sigma$ *for all* i.

Proof. Let $m \geq n$ be any two integers. We have the following commutative diagram with exact rows:

$$\begin{array}{ccccccccc} 0 & \longrightarrow & X_{\geq m} & \longrightarrow & X & \longrightarrow & X/X_{\geq m} & \longrightarrow & 0 \\ & & \downarrow & & \| & & \downarrow & & \\ 0 & \longrightarrow & X_{\geq n} & \longrightarrow & X & \longrightarrow & X/X_{\geq n} & \longrightarrow & 0 \end{array}$$

It gives rise to another commutative diagram with exact rows, one part of which looks as follows:

$$\begin{array}{ccccccc} \cdots & \longrightarrow & \underline{\mathrm{Ext}}^i_{\mathcal{M}}(X/X_{\geq m}, Y) & \longrightarrow & \underline{\mathrm{Ext}}^i_{\mathcal{M}}(X, Y) & \longrightarrow & \underline{\mathrm{Ext}}^i_{\mathcal{M}}(X_{\geq m}, Y) \\ & & \uparrow & & \| & & \uparrow \\ \cdots & \longrightarrow & \underline{\mathrm{Ext}}^i_{\mathcal{M}}(X/X_{\geq n}, Y) & \longrightarrow & \underline{\mathrm{Ext}}^i_{\mathcal{M}}(X, Y) & \longrightarrow & \underline{\mathrm{Ext}}^i_{\mathcal{M}}(X_{\geq n}, Y) \end{array}$$

$$\begin{array}{cccc} \longrightarrow & \underline{\mathrm{Ext}}^{i+1}_{M}(X/X_{\geq m}, Y) & \longrightarrow & \cdots \\ & \uparrow & & \\ \longrightarrow & \underline{\mathrm{Ext}}^{i+1}_{M}(X/X_{\geq n}, Y) & \longrightarrow & \cdots \end{array}$$

By Lemma 2.13 (b), the entries from the first and fourth columns are objects of $\overline{C}$. Since Theorem (A-5) holds in $\mathcal{M}$, we have that the inductive limit of rows by m is exact in $\mathcal{M}$:

$$\cdots \longrightarrow Z^i \longrightarrow \underline{\mathrm{Ext}}^i_{\mathcal{M}}(X, Y) \longrightarrow \mathrm{Ext}^i_{\mathcal{M}\Sigma^{-1}}(X, Y) \longrightarrow Z^{i+1} \longrightarrow \cdots$$

Here $Z^k \in \mathrm{Ob}\,\overline{C}$. Therefore the morphism in the middle belongs to $'\Sigma$.

PROOF OF THEOREM (A-6). The exact sequence $0 \to A_{\geq m} \to A \to A/A_{\geq m} \to 0$ gives rise to the isomorphism

$$\mathrm{Ext}^i_{\mathcal{M}}(A_{\geq m}, F) \cong \mathrm{Ext}^{i+1}_{\mathcal{M}}(A/A_{\geq m}, F) = 0$$

for $i \geq n > 0$. The assertion of the theorem follows now from Theorem 3.3 and Lemma 2.11.

§4. The Veronese equivalence

We now choose a natural number $d > 0$. Along with the graded algebra A we also consider the graded left Noetherian algebra $A^{(d)}$, where $(A^{(d)})_i = A_{d\cdot i}$. The categories of graded left A- and $A^{(d)}$-modules will be denoted $\mathcal{M}$ and $\mathcal{M}^{(d)}$ with the corresponding subcategories denoted by $\mathcal{M}^0$, C and $\overline{C}$. The symbol (d) will not be mentioned if it can be recovered from the context.

We define a pair of functors

$$V: \mathcal{M} \mapsto \mathcal{M}^{(d)}, \qquad T: \mathcal{M}^{(d)} \mapsto \mathcal{M},$$

where

$$V(M)_i = M_{d\cdot i}, \qquad T(N)_j = (A \otimes_{A^{(d)}} N)_j := \sum_{k\cdot d+l=j} A_l \otimes_{A^{(d)}} N_k .$$

Their action on morphisms is defined in the obvious way.

We remind the reader that $A = A_0\langle A_1 \rangle$.

LEMMA 4.1. *The functor V preserves subcategories $\mathcal{M}^0$, C, and $\overline{C}$.*

PROOF. If $M \in \mathrm{Ob}\mathcal{M}^0 \subset \mathrm{Ob}\mathcal{M}$ then there exists a k such that the module $M_{\geq k\cdot d}$ is generated by the component $M_{k\cdot d}$ and is also finitely generated. Since A is Noetherian it now follows that $V(M)_{\geq k}$ is finitely generated in $\mathcal{M}^{(d)}$ and therefore $V(M) \in \mathrm{Ob}\mathcal{M}^0 \subset \mathrm{Ob}\mathcal{M}^{(d)}$.

If $M \in \mathrm{Ob}\,\overline{C} \subset \mathrm{Ob}\mathcal{M}$ and $x \in V(M)$ then x can be viewed as an element of M (up to grading). By Corollary 2.3, we can find a k such that $A^{(d)}_{\geq k}\cdot x \subseteq A_{\geq k\cdot d}\cdot x = 0$ and therefore $V(M) \in \mathrm{Ob}\,\overline{C} \subset \mathrm{Ob}\mathcal{M}^{(d)}$.

Since $\mathrm{Ob}\,C = \mathrm{Ob}\,\overline{C} \cap \mathrm{Ob}\mathcal{M}^0$ we have that V preserves C.

LEMMA 4.2. *The functor T preserves subcategories $\mathcal{M}^0$, C, and $\overline{C}$.*

PROOF. If $N \in \mathrm{Ob}\mathcal{M}^0 \subset \mathrm{Ob}\mathcal{M}^{(d)}$ then we can find an n such that $N_{\geq n}$ is finitely generated. But $A = A_0\langle A_1 \rangle$ and therefore

$$T(N)_{\geq (n+1)\cdot d} = (A \otimes_{A^{(d)}} N)_{\geq (n+1)\cdot d} \subseteq A \otimes_{A^{(d)}} (N_{\geq n}).$$

The last module is finitely generated, therefore the first one is also finitely generated and $T(N) \in \mathrm{Ob}\mathscr{M}^0 \subset \mathrm{Ob}\mathscr{M}$.

Suppose now that $N \in \mathrm{Ob}\overline{C} \subset \mathrm{Ob}\mathscr{M}^{(d)}$ and $x \in T(N)$. Then $x = \sum a_i \otimes n_i$, where $a_i \in A$, $n_i \in N$. There exists a k such that $A^{(d)}_{\geq k} \cdot n_i = 0$ for all i and therefore

$$A_{\geq k \cdot d} \cdot x = 0.$$

It now follows from Corollary 2.3 that $T(N) \in \mathrm{Ob}\overline{C} \subset \mathrm{Ob}\mathscr{M}$.

The assertion about category C follows from the above.

COROLLARY 4.3. *The functor V is exact and*

$$\mathrm{Tor}_1^{A^{(d)}}(A, C') \in \mathrm{Ob}\overline{C}:$$

the functors T and V can be extended to the functors

$$T_1\colon \mathrm{Qco}(A^{(d)}) \mapsto \mathrm{Qco}(A); \qquad T_2\colon \mathrm{Coh}(A^{(d)}) \mapsto \mathrm{Coh}(A);$$
$$V_1\colon \mathrm{Qco}(A) \mapsto \mathrm{Qco}(A^{(d)}); \qquad V_2\colon \mathrm{Coh}(A) \mapsto \mathrm{Coh}(A^{(d)}).$$

THEOREM 4.4. *The functors T_1, V_1 and T_2, V_2 give rise to an equivalence of categories.*

PROOF. It is clear that

$$V \circ T = \mathrm{Id}_{\mathscr{M}^{(d)}}.$$

Suppose that $M \in \mathrm{Ob}\mathscr{M}$. Consider the following morphism in $\mathscr{M}$:

$$f\colon T \circ V(M) \to M, \qquad \sum a_i \otimes m_i \to \sum a_i \cdot m_i.$$

Let x be a homogeneous element of the kernel of f. Taking into account that $A = A_0\langle A_1\rangle$, we can write $x = \sum a_i \otimes m_i$, where $a_i \in A$ for all i, $0 \leq l \leq d$ and $m_i \in M_{k \cdot d}$ (since the degrees of m_i are multiples of d, we identify $V(M)$ with a subset of M and $A_{\geq j} = A \cdot A_j$ for all $j \geq 0$). We now have that

$$\begin{aligned} A_{\geq d-l} \cdot x = A \cdot A_{d-l} \cdot x = A \cdot A_{d-l} \cdot \sum a_i \otimes m_i \\ = A \cdot (1 \otimes A_{d-l} \sum a_i \cdot m_i) = 0 \end{aligned}$$

and it now follows from Corollary 2.3 that

$$\ker(f) \in \mathrm{Ob}\overline{C} \subset \mathrm{Ob}\mathscr{M}.$$

If $m \in M$ then $A_{\geq d} \cdot m = A \cdot A_d \cdot m \subset \mathrm{Im}(f)$, and therefore $\mathrm{Coker}(f) \in \mathrm{Ob}\overline{C} \subset \mathrm{Ob}\mathscr{M}$. This, along with Corollary 4.3, finishes the proof of the theorem. □

The author thanks Yu. I. Manin for posing the problem and for help during the work.

BIBLIOGRAPHY

1. J.-P. Serre, *Coherent algebraic bundles*, Ann. of Math. (2) **61** (1955), 197–278.
2. A. Connes, *Noncommutative differential geometry*, Inst. Hautes Études Sci. Publ. Math. **62** (1985), chapters I–II.
3. Yu. I. Manin, *Gauge field theory and complex geometry*, "Nauka", Moscow, 1984; English transl., Springer-Verlag, Berlin-New York, 1988.
4. M. M. Kapranov, *Derived category of coherent bundles on a quadric*, Funktsional Anal. i Prilozhen. **20** (1986), no. 2, 67–68; English transl. in Functional Anal. Appl. **20** (1986), no. 2.
5. R.-O. Buchweitz, *Maximal Cohen-Macaulay modules and Tate-cohomology over Gorenstein rings*, Preprint, 1987.
6. A. Grothendieck, *On some problems of homological algebra*, Tôhoku Math. J. (2) **9** (1957), 119–221.
7. J. Backelin and R. Froberg, *Koszul algebra, Veronese subrings and rings with linear resolutions*, Rev. Roumaine Math. Pures Appl. **30** (1985), no. 2, 85–97.
8. P. Gabriel and M. Zisman, *Calculus of fractions of homotopy theory*, Springer-Verlag, New York, 1967.
9. H. Bass, *Algebraic K-theory*, Benjamin, New York, 1968.
10. J. Milne, *Etale cohomology*, Princeton University Press, Princeton, NJ, 1980.

Translated by A. MARTSINKOVSKY

Amer. Math. Soc. Transl.
(2) Vol. **151**, 1992

Jordan and Lie Superalgebras Determined by a Poisson Algebra

I. L. KANTOR

Suppose U is a Poisson algebra, i.e., a (super)space with two operations: a commutative associative operation $(x, y) \mapsto xy$ and a Lie operation $(x, y) \mapsto \{x, y\}$, jointly satisfying the Leibniz identity. In this paper we introduce two functors, $\mathfrak{J}$ and $\mathfrak{L}$, which assign to a Poisson algebra U a Jordan superalgebra $\mathfrak{J}(U)$ and a Lie superalgebra $\mathfrak{L}(U)$.

Both superalgebras $\mathfrak{J}(U)$ and $\mathfrak{L}(U)$ are defined on the space $U \dotplus U^S$ of doubled dimension, where U^S is a duplicate of the space U with opposite parity of elements. Multiplication in the superalgebra $\mathfrak{J}(U)$ is defined by the formulas

$$x * y = xy, \quad x^S * y = (-1)^x x * y^S = (xy)^S, \quad x^S * y^S = (-1)^S\{x, y\} \tag{1}$$

and in the superalgebra $\mathfrak{L}(U)$ by the formulas

$$[x, y] = \{x, y\}, \quad [x^S, y] = (-1)^x[x, y^S] = \{x, y\}^S, \quad [x^S, y^S] = (-1)^x xy. \tag{2}$$

It turns out that the superalgebra $\mathfrak{J}(U)$ defined by the formulas (1) remains a Jordan algebra under weaker conditions on the commutative operation xy of the Poisson algebra: associativity is replaced by the Jordan property, but we also need a fourth-degree identity connecting the associator with the Lie operation $\{x, y\}$. It is natural to call such algebras Poisson-Jordan algebras.

Analogously, the superalgebra $\mathfrak{L}(U)$ defined by the formulas (2) remains a Lie algebra if we simply omit the associativity of the commutative operation. Algebras satisfying only the Jacobi and Leibniz identities will be called generalized Poisson algebras. Thus the functors $\mathfrak{J}$ and $\mathfrak{L}$ are defined on Poisson-Jordan algebras and generalized Poisson algebras, respectively.

1991 *Mathematics Subject Classification.* Primary 17A70, 17B70, 17C70.

An important property of the functor $\mathfrak{J}$ is the following: The Jordan superalgebra $\mathfrak{J}(U)$ is simple if and only if the Poisson algebra U is simple. Let Γ_n be the Grassman algebra with (odd) generators $\xi_1, \xi_2, \dots, \xi_n$. Let H_n denote the simple Poisson algebra on Γ_n with Lie operation

$$\{x, y\} = (-1)^x \sum \frac{\partial x}{\partial \xi_i} \frac{\partial y}{\partial \xi_i}.$$

The series of simple Jordan superalgebras $\mathfrak{J}(H_n)$ was omitted from the classification obtained in [1].

On the other hand, the indicated property of the functor $\mathfrak{J}$ enables us to obtain, starting from a classification of simple Jordan superalgebras, a classification of simple Poisson-Jordan algebras. This classification, in which Poisson algebras are defined to within multiplication of one of the operations by a scalar factor, includes in addition to the series of ordinary Poisson algebras H_n the two other series $A_{m,n}$ and Q_n.

The functor $\mathfrak{L}$ enables us to obtain an analogous, but weaker, result. In this case, in addition to the absence of ordinary ideals we also require the absence of weak ideals. The classification of generalized Poisson algebras without weak ideals, to within multiplication of one of the operations by a scalar factor, contains only one series of algebras $\widetilde{A}_{m,n}$, which is a modification of the series of algebras $A_{m,n}$.

The first two sections are preliminary in nature. The construction of a Jordan trilinear operation in §2, which is perhaps of independent interest, is heuristic for the construction of the superalgebras $\mathfrak{J}(U)$ and $\mathfrak{L}(U)$. In §3 we introduce the functor $\mathfrak{J}$ for the case of ordinary Poisson algebras and study its properties. In §4 we correct the classification of simple finite-dimensional Jordan superalgebras. In §5 the action of the functor $\mathfrak{J}$ is extended to Poisson-Jordan algebras. In §6 we consider the functor $\mathfrak{L}$ and generalized Poisson brackets.

In this paper the term "Poisson algebra" is applied both to an ordinary "even" space and to a superspace. Analogously, the prefix "super" is sometimes omitted in other terms, since it is natural to assume, for example, that a Lie superalgebra is a Lie algebra defined on a superspace.

The author thanks E. I. Zel′manov for his unfailing interest in this research and for useful discussions.

§1. Definitions and some identities

DEFINITION 1.1. By a Poisson algebra we mean a linear space with two bilinear operations: a commutative associative operation $(a, b) \mapsto ab$ and a Lie operation $(a, b) \mapsto \{a, b\}$, jointly satisfying the Leibniz "rule", i.e., satisfying the following identities:

$$ab = (-1)^{ab} ba, \text{ (}^1\text{)} \tag{1.1}$$

$$(ab)c = a(bc), \tag{1.2}$$

$$\{a, b\} = -(-1)^{ab}\{b, a\}, \tag{1.3}$$

$$\{a, \{b, c\}\} = \{\{a, b\}, c\} - (-1)^{bc}\{\{a, c\}, b\}, \tag{1.4}$$

$$\{ab, c\} = a\{b, c\} + (-1)^{ab} b\{a, c\}. \tag{1.5}$$

From these identities, with the aid of (1.1) and (1.3), we can easily obtain other analogous relations. In particular,

$$\{\{a, b\}, c\} = \{a, \{b, c\}\} - (-1)^{ab}\{b, \{a, c\}\},$$

$$\{ab, c\} = a\{b, c\} + (-1)^{bc}\{a, c\}b,$$

$$\{a, bc\} = \{a, b\}c + (-1)^{ab} b\{a, c\},$$

$$\{a, bc\} = \{a, b\}c + (-1)^{bc}\{a, c\}b.$$

The important identity

$$\{a, bc\} = \{ab, c\} + (-1)^{bc}\{ac, b\} \tag{1.6}$$

can be proved by applying the Leibniz rule to all terms.

Note that if the associative operation has a unity e, then e belongs to the center of the Lie algebra:

$$\{e, x\} = 0, \quad \forall x. \tag{1.7}$$

This follows from the Leibniz identity with $a = b = e$.

An important example of a Poisson algebra on an even space is the space of functions of $2n$ variables $x_1, x_2, \dots, x_n, x_1', x_2', \dots, x_n'$ with the associative operation of multiplication of functions and the Lie operation

$$\{f, g\} = \sum_{i=1}^{n} \left(\frac{\partial f}{\partial x_i} \frac{\partial g}{\partial x_i'} - \frac{\partial f}{\partial x_i'} \frac{\partial g}{\partial x_i} \right). \tag{1.8}$$

Another example of a Poisson algebra on an even space is that of the Poisson-Lie brackets on the space of polynomials in n variables $x_1, x_2, \dots, x_n$ with the associative operation of multiplication of polynomials and the Lie operation defined for the generators $x_1, x_2, \dots, x_n$ by the formula

$$\{x_i, x_j\} = \sum_{k=1}^{n} C_{ij}^k x_k, \tag{1.9}$$

where the C_{ij}^k are the structure constants of a fixed Lie algebra on an n-dimensional space. For the other polynomials the operation is defined by means of (1.9) and the Leibniz identity.

(1) Here and henceforth, $(-1)^{ab} \equiv (-1)^{\deg a \cdot \deg b}$.

An example of a Poisson algebra on a space containing odd elements is the space of the Grassmann algebra Γ_n with odd generators $\xi_1, \xi_2, \dots, \xi_n$ and the Lie operation

$$\{f, g\} = (-1)^f \sum_{i=1}^{n} \frac{\partial f}{\partial \xi_i} \frac{\partial g}{\partial \xi_i}. \tag{1.10}$$

DEFINITION 1.2. By a Jordan (super)algebra we mean a linear space with a bilinear operation $(a, b) \mapsto a * b$ satisfying the two identities

$$a * b = (-1)^{ab} b * a, \tag{1.11}$$

$$[L_{a*b}, L_c] = [L_a, L_{b*c}] + (-1)^{ab}[L_b, L_{a*c}], \tag{1.12}$$

where L_a is the left shift linear operator $L_a(x) = a * x$ and square brackets denote the commutator of two linear operators: $[A, B] = AB - (-1)^{AB} BA$.

We mention one important consequence of (1.11) and (1.12):

$$[[L_a, L_b], L_c] = L_{a*(b*c)} - (-1)^{ab} b * (a * c), \tag{1.13}$$

which means that the linear transformations $[L_a, L_b]$ are derivations of the original Jordan algebra.

An important example of a Jordan algebra is the algebra $A_{m,n}$ defined on the space of square matrices of order $m+n$, where m is the number of odd variables and n is the number of even variables, with the operation

$$A * B = AB + (-1)^{AB} BA. \tag{1.14}$$

With some exceptions, finite-dimensional Jordan algebras are subalgebras of the $A_{m,n}$.

DEFINITION 1.3. By a Jordan trilinear operation we mean a linear space with a trilinear operation $(a, b, c) \mapsto \langle a, b, c\rangle$ satisfying the identities

$$\langle a, b, c\rangle = (-1)^{bc}\langle a, c, b\rangle \tag{1.15}$$

$$\begin{aligned} \langle a, b, \langle u, v, w\rangle\rangle = &-(-1)^{ab}\langle\langle b, a, u\rangle, v, w\rangle \\ &+ (-1)^{(a+b)u}\langle u, \langle a, b, v\rangle, w\rangle \\ &+ (-1)^{(a+b)(u+v)}\langle u, v, \langle a, b, w\rangle\rangle. \end{aligned} \tag{1.16}$$

REMARK. In contrast to the usual definition, here the first and second arguments have changed roles. This definition is more convenient for technical reasons, especially in the super case.

Condition (1.16) means that the linear transformations $L_{ab}(x) = \langle a, b, x\rangle$ are "almost" derivations: the first term on the right-hand side should have been $\langle\langle a, b, u\rangle, v, w\rangle$.

A typical example of a finite-dimensional Jordan trilinear operation is the space of matrices of order $m + n$, where m and n are the number of odd and even variables, with the operation

$$\langle A, B, C\rangle = (-1)^{AB} BAC + (-1)^{AB+(A+B)C} CAB. \tag{1.17}$$

An important property of Jordan trilinear operations is included in the next theorem.

THEOREM 1.1. *Suppose a is an even element and ξ is an odd element. Then*

1) *the bilinear operation $x*y = \langle a, x, y\rangle$ defines on the original space the structure of a Jordan (super)algebra;*

2) *the bilinear operation $[x*y] = (-1)^x\langle \xi, x, y\rangle$ defines on the original space, the parity of whose elements is changed into the opposite, the structure of a Lie (super)algebra.*

The proof of the first assertion can be carried out, exactly as in the even case, by constructing a graded Lie (super)algebra of the form

$$U_{-1} \dotplus U_0 \dotplus U_1 \tag{1.18}$$

with an involutive automorphism σ such that $\sigma(U_i) = U_{-i}$. The original Jordan trilinear operation can be represented in the form

$$\langle a, b, c\rangle = [[\sigma(a), b], c], \quad \forall a, b, c \in U_{-1} \tag{1.19}$$

(see [2]). The technical changes consist only in the application of the rule of signs.

The proof of the second assertion, using the construction of the algebra (1.18), can be found in [3], where it is proved that to each odd element $a \in U_1$ there corresponds a Lie (super)algebra on the space U_{-1}^S with operation $\{x, y\} = (-1)^x[[a, x], y]$.

§2. Construction of a Jordan trilinear operation from a Poisson algebra

For each linear space U we denote by U^S the linear space that is a duplicate of U, but with the parity of elements changed into the opposite.

THEOREM 2.1. *Suppose U is the space of a Poisson algebra. Then the trilinear operation*

$$\langle x, y, z\rangle = (-1)^x(\{xy, z\} - \{x, y\}z) \tag{2.1}$$

defines on the space U^S the structure of a Jordan trilinear operation.

The proof will be carried out by direct verification. It is important to observe that the elements in identity (2.1) are elements of the space U^S, and, to apply the Jacobi and Leibniz identities, these elements must be viewed as elements of the space U. Therefore, in passing from the left- to the right-hand side in the following equalities, make the replacement $a^S = a + 1$ in the exponent of (-1).

We write down separately all four terms of identity (2.1):

$$\langle a, b, \langle u, v, w\rangle\rangle = (-1)^{b+v}(\underset{\sim\sim\sim}{\{ab, \{uv, w\}\}} - \underset{\approx\approx\approx}{\{ab, \{u, v\}w\}} - \{a, b\}\{uv, w\} + \underline{\{a, b\}\{u, v\}w}),$$

$$\begin{aligned} &- (-1)^{a^S b^S} \langle\langle b, a, u\rangle, v, w\rangle \\ &\quad = -(-1)^{ab+v+b+1}(\utilde{\{\{ba, u\}v, w\}} - \underline{\underline{\{\{b, a\}uv, w\}}} \\ &\qquad\qquad - \utilde{\{\{ba, u\}v\}w} + \{\{ba, u\}v, w\}), \end{aligned}$$

$$\begin{aligned} &(-1)^{(a^S+b^S)u^S} \langle u, \langle a, b, v\rangle, w\rangle \\ &\quad = (-1)^{(a+b)u+v+b}(\utilde{\{u, \{ab, v\}, w\}} - \underline{\{u\{a, b\}v, w\}} \\ &\qquad\qquad - \utilde{\{u, \{ab, v\}\}w} + \{u, \{a, b\}v\}w, \end{aligned}$$

$$\begin{aligned} &(-1)^{(a^S+b^S)(u^S+v^S)} \langle u, v, \langle a, b, w\rangle\rangle \\ &\quad = (-1)^{(a+b)(u+v)+v+b}(\utilde{\{uv, \{ab, w\}\}} - \{uv, \{a, b\}w\} \\ &\qquad\qquad - \utilde{\{u, v\}\{ab, w\}} + \underline{\{u, v\}\{a, b\}w}). \end{aligned}$$

If we substitute these expressions into (2.1), then the terms underlined with straight lines cancel directly, the terms underlined with wavy lines cancel by means of the Leibniz and Jacobi identities, and the nonunderlined terms cancel by means of the Leibniz identity and identity (1.6). The theorem is proved.

Theorems 2.1 and 1.1 have the following consequences.

THEOREM 2.2. *The operations*

$$[x, y]_a = \{ax, y\} - \{a, x\}y, \quad \deg a = 0,$$

form on the space of the Poisson algebra U *a linear bundle of Lie algebras jointly satisfying the additional identity*

$$[x, [y, z]_b]_a - [[x, y]_a, z]_b + (-1)^{yz}[[x, z]_a, y]_b = -[y, z])_{[a, b]_x} \tag{2.2}$$

REMARK. Identity (2.2) is a generalization of the Jacobi identity, with which it agrees when $a = b$.

Note that if the associative operation has a unity e, then it follows from (2.1) that the bundle of the Lie algebras includes the original Lie operation $[x, y]_e = \{x, y\}$.

THEOREM 2.3. *The operations*

$$(x * y)_a = (-1)^x(\{ax, y\} - \{a, x\}y), \quad \deg a = 1$$

form on the space U^S, *corresponding to the space of the Poisson algebra* U, *a linear bundle of Jordan algebras jointly satisfying the additional identity*

$$\begin{aligned} &(x * (y * z)_b)_a - (-1)^x((x * y)_a * z)_b - (-1)^{xy}(y * (x * z)_a)_b \\ &\quad = (-1)^x(y * z)_{(a*b)_x}. \end{aligned} \tag{2.3}$$

§3. A correspondence between Poisson algebras and Jordan superalgebras

The Jordan trilinear operation obtained in §2 does not enable us, in general, to assign to a Poisson algebra a completely determined Jordan algebra. Such a construction is given in Theorem 3.1, but it is necessary to double the dimension of the original space. Note that this construction is a consequence of Theorem 2.1 in the case where the Poisson algebra has a unity. Theorem 3.1 is proved without this restriction.

Suppose U is the space of a Poisson algebra. We consider the direct sum of spaces $U \dotplus U^S$ and define on it an operation $*$ by the formulas

$$\begin{gathered} a * b = ab, \qquad a^S * b = (-1)^a a * b^S = (ab)^S, \\ a^S * b^S = (-1)^a \{a, b\}, \qquad a, b \in U, \quad a^S, b^S \in U^S. \end{gathered} \tag{3.1}$$

Recall that a^S denotes the copy of the element $a \in U$ in the space U^S.

DEFINITION 3.1. By the algebra $\mathfrak{J}(U)$ we mean the space $U \dotplus U^S$ with the operation defined by (3.1).

THEOREM 3.1. *The algebra $\mathfrak{J}(U)$ is a Jordan (super)algebra.*

PROOF. We must verify identities (1.11) and (1.12). Identity (1.11) is a consequence of the following calculations:

$$\begin{aligned} a^S * b &= (ab)^S = (-1)^{ab}(ba)^S = (-1)^{ab+b} b * a^S = (-1)^{a^S b} b * a, \\ a^S * b^S &= (-1)^a \{a, b\} = -(-1)^{a+ab}\{b, a\} \\ &= (-1)^{a^S b^S + b}\{b, a\} = (-1)^{a^S b^S} b^S * a^S. \end{aligned}$$

To prove (1.12) we denote by P_a and Q_a the linear "shift" operations on U:

$$P_a(x) = ax, \qquad Q_a(x) = \{a, x\}.$$

Using properties of a Poisson algebra, we can easily verify the following relations:

$$\begin{gathered} [P_a, P_b] = 0, \quad [P_a, Q_b] = P_{\{a,b\}}, \quad [Q_a, Q_b] = Q_{\{a,b\}}, \\ Q_a P_b + (-1)^{ab} Q_b P_a = Q_{ab}. \end{gathered} \tag{3.2}$$

It follows from the definition of the operation $*$ that the left shifts in the algebra $\mathfrak{J}(U)$ can be represented by the "second-order" matrices

$$L_a = \begin{pmatrix} P_a & \\ & (-1)^a P_a \end{pmatrix}, \qquad L_{a^S} = \begin{pmatrix} & (-1)^a Q_a \\ P_a & \end{pmatrix}. \tag{3.3}$$

Using (3.2), we can easily verify the relations

$$\begin{gathered} [L_a, L_b] = 0, \qquad [L_a, L_{b^S}] = \begin{pmatrix} 0 & (-1)^b P_{\{a,b\}} \\ 0 & 0 \end{pmatrix}, \\ [L_{a^S}, L_{b^S}] = \begin{pmatrix} (-1)^a P_{ab} & 0 \\ 0 & (-1)^b P_{ab} \end{pmatrix}. \end{gathered} \tag{3.4}$$

Let us now consider the various cases in identity (1.12). If $a, b, c \in U$, then it follows from (3.2) that all three terms are equal to zero. If $a \in U$, $b = u^S$, $c = v \in U^S$, then with the aid of (3.4) the verification reduces to verifying the relation

$$P_{\{ab, u\}} = P_{\{a, bu\}} + (-1)^{ab} P_{\{b, au\}}, \tag{3.5}$$

which is a consequence of (1.6). If $a \in U$, $b = u^S$, $c = v^S \in U^S$, then, according to (3.4), we must verify that

$$Q_{u(va)} - (-1)^{uv} Q_{v(ua)} = 0. \tag{3.6}$$

This follows from the associativity of the operation xy. Finally, if $a = u^S$, $b = v^S$, $c = w^S \in U^S$, then, by (3.4), the verification reduces to verifying that

$$P_{\{\{u, v\}, w\}} = -(-1)^{(v+w)u} P_{\{\{v, w\}, u\}} - (-1)^{vw} P_{\{\{u, w\}, v\}},$$

which follows from the Jacobi identity (1.4). Thus in all cases the relation (1.12) holds. The theorem is proved.

In order to make the correspondence $U \mapsto \mathfrak{J}(U)$ invertible we consider the following class of Jordan algebras.

DEFINITION 3.2. Suppose σ is an involutive automorphism of a Jordan algebra A such that for a decomposition $A = A_1 \dotplus A_{-1}$ of A into σ-invariant subspaces there exists a one-to-one odd linear mapping $S: A_1 \to A_{-1}$.

The Jordan algebra A will be called an associative Jordan algebra of the form $H(\sigma, S)$ if

1) the subalgebra A_1 is associative,

2) $S(a) * b = S(a * b)$, $\forall a, b \in A_1$,

3) the Jordan algebra identity (1.12) remains valid if along with the other three elements we substitute a symbolic element S such that

$$S * x = S(x), \quad \forall x \in A_1, \qquad S * y = 0, \quad \forall y \in A_1.$$

REMARK. If a Jordan algebra has a unity e, then the third condition is unnecessary, since such an element S is, in view of condition 2), the element $S = S(e)$.

DEFINITION 3.3. Suppose A is an associative Jordan algebra of the form $H(\sigma, S)$. By the algebra $P(A)$ we mean the algebraic system on the subspace A_1 with the two operations

$$xy = x * y, \quad \{x, y\} = (-1)^x x^S * y^S, \quad \forall x, y \in A_1. \tag{3.7}$$

THEOREM 3.2. *Suppose A is an associative Jordan algebra of the form $H(\sigma, S)$. Then $P(A)$ is a Poisson algebra.*

PROOF. Since the operation xy is commutative and associative, we must verify the skew-symmetry of the operation $\{x, y\}$ and the Jacobi and Leibniz identities.

The skew-symmetry of the operations $\{\ ,\ \}$ can be proved in a simple way:

$$\{x, y\} = (-1)^x(S(x) * S(y)) = (-1)^{x+(x+1)(y+1)}S(y) * S(x)$$
$$= -(-1)^{xy}(-1)^y S(y) * S(x) = -(-1)^{xy}\{y, x\}.$$

To prove the Jacobi identity we substitute $a = S(u)$, $b = S(v)$, $c = S(w)$ in (1.12), where $u, v, w \in A_1$ and as the argument we take an element S. Then, using $x * S = (-1)^x S * x$, we obtain

$$-(-1)^{(u+v)(w+1)}(-1)^{v+w}\{w, \{u, v\}\}$$
$$= (-1)^{u+w}\{u, \{v, w\}\}$$
$$+ (-1)^{(u+1)(v+1)}(-1)^{v+w}\{v, \{u, w\}\}.$$

Now using skew-symmetry and multiplying by $(-1)^{u+w}$, we obtain the Jacobi identity

$$\{\{u, v\}, w\} = \{u, \{v, w\}\} - (-1)^{uv}\{v, \{u, w\}\}.$$

To prove the Leibniz identity we substitute $a = S(u)$, $b = S(v)$, $c = S$ in (1.12) and as the argument we take $S(w)$, where $u, v, w \in A_1$. Then we obtain

$$S(\{uv, w\}) = S(u)\{v, w\} - (-1)^{(u+1)(v+1)+u+v}S(v)\{u, w\}.$$

Using condition 2) and the fact that the mapping S is one-to-one, we obtain the Leibniz identity

$$\{uv, w\} = u\{v, w\} + (-1)^{uv}v\{u, w\}.$$

The theorem is proved.

From Theorems 3.1 and 3.2 there follows

THEOREM 3.3. *The mappings $\mathfrak{J}$ and P are mutually inverse and establish a one-to-one correspondence between the Poisson algebras and the associative-Jordan algebras of the form $H(\sigma, S)$. This correspondence is functorial.*

PROOF. An elementary comparison of formulas (3.1) and (3.7) shows for a Poisson algebra U and an associative-Jordan algebra A of the form $H(\sigma, S)$ we have

$$P\mathfrak{J}(U) = U, \qquad \mathfrak{J}P(A) = A, \tag{3.8}$$

i.e., $\mathfrak{J}$ and P are mutually inverse correspondences.

By an S-ideal of a Jordan algebra $A \in H(\sigma, S)$ we mean an ideal of the form

$$\mathscr{D}^* = \mathscr{D} \dotplus S(\mathscr{D}),$$

where $\mathscr{D}$ is an ideal of A_1. The factor algebra $A/\mathscr{D}^*$ is an algebra of the form $H(\sigma, S)$; the homomorphism $A \to A/\mathscr{D}^*$ will be called an S-homomorphism.

Consider two categories: the category P of Poisson algebras with homomorphisms as morphisms, and the category H of associative Jordan algebras of the form $H(\sigma, S)$ with S-homomorphisms as morphisms.

It is obvious that the diagrams

$$\begin{array}{ccccccc} U & \xrightarrow{\mathfrak{J}} & \mathfrak{J}(U) & \quad & A & \xrightarrow{P} & P(A) \\ \downarrow{\scriptstyle\varphi} & & \downarrow{\scriptstyle\varphi'} & & \downarrow{\scriptstyle\psi} & & \downarrow{\scriptstyle\psi'} \\ U' & \xrightarrow{\mathfrak{J}} & \mathfrak{J}(U') & & A' & \xrightarrow{P} & P(A') \end{array}$$

are commutative, since an ideal $\mathscr{D}$ of the Poisson bracket $P(A)$ is uniquely determined by an S-ideal of A and an ideal $\mathscr{D}^*$ of $\mathfrak{J}(U)$ is uniquely determined by an ideal of U.

DEFINITION 3.4. A Poisson algebra will be called simple if it contains no nontrivial ideals, i.e., no subspaces $\mathscr{D}$ such that

$$\{x, \mathscr{D}\} \subset \mathscr{D}, \qquad x \cdot \mathscr{D} \subset \mathscr{D}, \qquad \forall x.$$

It follows from Theorem 3.3 that if a Poisson algebra is simple, then the algebra $\mathfrak{J}(U)$ contains so S-ideals.

We will prove a stronger assertion. We first make the following definition.

DEFINITION 3.5. By a quasi-ideal of a Poisson algebra we mean a subspace $\mathscr{D}$ such that

$$x\mathscr{D} \subset \mathscr{D}, \qquad \{x, \mathscr{D}\}y \subset \mathscr{D}, \qquad \forall x, y. \tag{3.9}$$

REMARK. The introduction of the concept of a quasi-ideal was motivated by the desire to consider Poisson algebras without a unity. Obviously if there is a unity e, then by substituting $y = e$ in (3.9) we see that the concepts of ideal and quasi-ideal agree.

LEMMA 3.1. *A simple Poisson algebra contains no quasi-ideals.*

PROOF. Suppose $\mathscr{D}$ is a quasi-ideal. Let $\mathscr{D}_1$ denote a maximal subspace such that

$$x\mathscr{D}_1 \subset \mathscr{D}. \tag{3.10}$$

We will show that $\mathscr{D}_1$ is an ideal of the Poisson algebra. Note first that for a simple Poisson algebra U the subspace $U' = U \cdot U$ is equal to the whole space. Indeed, it follows from the Leibniz identity that U' is an ideal not only of the commutative operation, but also of the Lie operation. It follows from $U' = U$ that $\mathscr{D}_1$ is a proper subspace of U.

From the definition of a quasi-ideal and (3.10) we see that

$$\{x, \mathscr{D}\} \subset \mathscr{D}_1. \tag{3.11}$$

Now suppose $a \in \mathscr{D}_1$ in identity (1.6). Then, using (3.10) and (3.11), we obtain

$$\{\mathscr{D}_1, xy\} \subset \mathscr{D}_1, \qquad \forall x, y.$$

Since $UU \subset U$, this means that

$$\{\mathscr{D}_1, U\} \subset \mathscr{D}_1.$$

In view of (3.10) and the relation $\mathscr{D} \subset \mathscr{D}_1$, it follows that $\mathscr{D}_1$ is a proper ideal of U, which contradicts our hypothesis.

THEOREM 3.4. *Suppose U is a Poisson algebra with nonzero operations xy and $\{x, y\}$. Then the Jordan algebra $\mathfrak{J}(U)$ is simple if and only if the Poisson algebra U is simple.*

PROOF. The "only if" part follows from the definition of multiplication in $\mathfrak{J}(U)$. The space $\mathscr{D} \dotplus \mathscr{D}^S$ is an ideal of $\mathfrak{J}(U)$ if $\mathscr{D}$ is an ideal of U.

To prove the converse we will use Lemma 3.1 repeatedly. Consider any ideal $\mathscr{D}$ of $\mathfrak{J}(U)$ and let $\mathscr{D}_1, \mathscr{D}_2^S$ denote its projections on the subspaces U, U^S. Multiplying $\mathscr{D}$ by arbitrary elements $a \in U$ and $a^S \in U^S$, we obtain (see (3.1)) the relations

$$a\mathscr{D}_1 \subset \mathscr{D}_1, \quad a\mathscr{D}_2 \subset \mathscr{D}_2, \quad a\mathscr{D}_1 \subset \mathscr{D}_2, \quad \{a, \mathscr{D}_2\} \subset \mathscr{D}_1. \tag{3.12}$$

It follows that $\mathscr{D}_1 \cap \mathscr{D}_2$ is a quasi-ideal of the Poisson algebra U. Thus there are two possibilities: 1) $\mathscr{D}_1 \cap \mathscr{D}_2 = 0$; 2) $\mathscr{D}_1 \cap \mathscr{D}_2 = U$.

Consider the first. It follows from (3.12) that $\mathscr{D}_1$ is the annihilator of the multiplication xy. The Leibniz identity

$$\{x, A\}y = \{x, Ay\} - \{x, y\}A$$

shows that the space A of annihilators of the operation xy is an ideal of U. Since U is a simple algebra and the multiplication xy is nonzero, we have $A = 0$, hence $\mathscr{D}_1 = 0$. But then $\mathscr{D}_2$ is an ideal of U. If $\mathscr{D}_2 \neq 0$, then $\mathscr{D}_2 = U$, which is also impossible, since the multiplication $\{x, y\}$ is nonzero.

Consider the second possibility, $\mathscr{D}_1 \cap \mathscr{D}_2 = U$, and distinguish in the ideal $\mathscr{D}$ a maximal subspace T containing the pairs $t \in U$ and $t^S \in U^S$. The intersection T_1 of the subspaces T and U is a quasi-ideal of U. Indeed, it follows from

$$x * T_1 = x \cdot T_1 \subset U, \qquad x^S * T_1 = (x \cdot T_1)^S \subset U^S, \quad \forall x$$

that $XT_1 \subset T_1$, $\forall x$, and from

$$y * (x^S * T_1^S) = y\{xT_1\} \subset U, \qquad y^S * (x^S * T_1^S) = (y\{x, T_1\})^S \subset U^S, \quad \forall x, y$$

that $y\{xT_1\} \subset T_1$, $\forall x, y$, i.e., T_1 is a quasi-ideal. Thus $T_1 = 0$.

It now follows from $T_1 = 0$ and $\mathscr{D}_1 \cap \mathscr{D}_2 = U$ that the elements of $\mathscr{D}$ have the form $a + (f(a))^S$, where $a \in U$, $(f(a))^S \in U^S$, and f is a one-to-one odd linear mapping of U into itself. Multiplying the elements of $\mathscr{D}$ on the right by $b \in U$, we obtain the relation

$$f(ab) = f(a)b. \tag{3.13}$$

Multiplying the elements of $\mathscr{D}$ on the right by $b^S \in U^S$, we obtain

$$f(\{f(a), b\}) = -ab. \tag{3.14}$$

Using (3.13), (3.14) and transforming $f^2(\{a, b\})$, we see that

$$f^2(\{a, b\}) = f(f(\{f(f^{-1}(a)), b\})) = -f(f^{-1}(a)b) = -ab,$$

i.e.,

$$f^2(\{a, b\}) = -ab. \tag{3.15}$$

But the last equality is impossible, since the skew-symmetric bilinear operation $f^2(\{a, b\})$ cannot be equal to the nonzero symmetric operation $-ab$. The theorem is proved.

§4. On the classification of simple Jordan superalgebras and the classification of simple Poisson algebras

In §1 we considered the Poisson algebra defined on the space of the Grassmann algebra Γ_n with odd generators $\xi_1, \xi_2, \dots, \xi_n$ and Lie operation

$$\{f, g\} = (-1)^f \sum_{i=1}^{n} \frac{\partial f}{\partial \xi_i} \frac{\partial g}{\partial \xi_i}. \tag{4.1}$$

This Poisson algebra is simple. Indeed, suppose the element a belongs to the ideal $\mathscr{D}$ and among the nonzero components of a the monomial $\xi_{i_1}\xi_{i_2}\cdots\xi_{i_k}$ has the largest number of factors (there are perhaps other components with the same number of factors that differ by at least one factor). Then it follows from (4.1) that the element $\{\xi_{i_k}, \{\xi_{i_{k-1}}, \dots, \{\xi_{i_2}\{\xi_{i_1}, a\}\}\cdots\}\}$ of the ideal $\mathscr{D}$ is collinear with the unit 1 of the Grassmann algebra, hence $\mathscr{D}$ coincides with the Grassmann algebra Γ_n.

If we denote this Poisson algebra by H_n, we see from Theorem 3.4 that the Jordan superalgebra $\mathfrak{J}(H_n)$ is simple. Let us denote this Jordan superalgebra by H_{n+1}. Recall that the Jordan algebra H_{n+1} is defined on the direct sum $\Gamma_n \dotplus \Gamma_n^S$, where Γ_n^S is a duplicate of the Grassmann algebra Γ_n, but the elements of Γ_n^S have the opposite parity. Multiplication in the superalgebra $H_{n+1} = \mathfrak{J}(H_n)$ is defined by the formulas

$$\begin{gathered} a_1 * a_2 = a_1 \cdot a_2, \qquad a^S * b = (-1)^a a * b^S = (a \cdot b)^S, \\ b_1^S * b_2^S = (-1)^{b_1}\{b_1, b_2\}, \end{gathered} \tag{4.2}$$

where $\cdot$ is the multiplication in the Grassmann algebra and the Lie operation $\{\ ,\ \}$ is defined by (4.1).

This simple Jordan superalgebra was omitted from the classification of Kac [1] (the omission is not due to a defect in the proof; apparently a grading of the form $U_{-1} \dotplus U_0 + U_1$ of the Lie superalgebra H_n was accidentally omitted, even though it is present among the gradings of general form found in [1]). Let us take note of two other inaccuracies in the classification.

A. The multiplication in the algebra $F(4)$ contains misprints. The correct multiplication table in $F(4)$ in the basis $e_1, e_2, \dots, e_6, \eta_1, \dots, \eta_4$ is the following:

$$e_1 * e_i = e_i \ (i = 1, 2, \dots, 5), \quad e_2^2 = e_3^2 = e_1, \quad e_4^2 = e_5^2 = -e_1,$$
$$e_6^2 = e_6, \quad e_1 * n_j = e_6 * \eta_j = \tfrac{1}{2}\eta_j \quad (j = 1, 2, 3, 4)$$

	e_2	e_3	e_4	e_5
η_1	$\frac{1}{2}\eta_3$	$\frac{1}{2}\eta_4$	$-\frac{1}{2}\eta_3$	$-\frac{1}{2}\eta_4$
η_2	$-\frac{1}{2}\eta_4$	$\frac{1}{2}\eta_3$	$-\frac{1}{2}\eta_4$	$\frac{1}{2}\eta_3$
η_3	$\frac{1}{2}\eta_1$	$\frac{1}{2}\eta_2$	$\frac{1}{2}\eta_1$	$-\frac{1}{2}\eta_2$
η_4	$-\frac{1}{2}\eta_2$	$\frac{1}{2}\eta_1$	$\frac{1}{2}\eta_2$	$\frac{1}{2}\eta_1$

$$\eta_1 * \eta_2 = e_1 - 3e_6, \quad \eta_1 * \eta_3 = e_3 + e_5, \quad \eta_1 * \eta_4 = -e_2 - e_4,$$
$$\eta_2 * \eta_3 = -e_2 + e_4, \quad \eta_2 * \eta_4 = -e_3 + e_5, \quad \eta_3 * \eta_4 = -e_1 + 3e_6.$$

The other products are equal to zero.

The indicated multiplication is given in a basis different from the basis in [1]. If we pass to the basis

$$a_1 = e_1, \quad a_6 = e_6, \quad a_2 = \frac{e_2 + e_4}{\sqrt{2}}, \quad a_3 = \frac{e_2 - e_4}{\sqrt{2}}, \quad a_4 = \frac{e_3 + e_5}{\sqrt{2}},$$
$$a_5 = \frac{e_3 - e_5}{\sqrt{2}}, \quad \xi_1 = \frac{1}{\sqrt{2}}\eta_1, \quad \xi_2 = -\eta_4, \quad \xi_3 = \eta_3, \quad \xi_4 = \sqrt{2}\eta_2,$$

then the correct table differs from that of [1] in the following products:

$$a_4 * \xi_2 = -\xi_1, \quad a_5 * \xi_1 = -\tfrac{1}{2}\xi_2, \quad \xi_1 * \xi_4 = e_1 - 3e_6,$$
$$\xi_2 * \xi_3 = -e_1 + 3e_6, \quad \xi_2 * \xi_4 = -2a_5, \quad \xi_3 * \xi_4 = 2a_3.$$

This correction appears in [3] and [4].

B. The nonisomorphic algebras in the series $\mathscr{D}_t$ are as follows. They are the algebras with basis $1, b, \xi, \eta$ and multiplication

$$b^2 = 1, \quad b\xi = b\eta = 0, \quad \xi\eta = 1 + bt \qquad (t \geq 0, \ t \neq 1). \tag{4.3}$$

The list of algebras in $\mathscr{D}_t$ given in [1] contains isomorphic algebras. Indeed, multiplication in algebras in the series $\mathscr{D}_{t'}$ was given in [1] in another basis $1, a, \xi', \eta$ by the table

$$a^2 = 2a, \quad a\xi' = \xi', \quad a\eta = \eta, \quad \xi'\eta = 1 + at'. \tag{4.4}$$

Consider first the case $t' = -1$. Passing to the basis

$$1, \quad b = 1 - a, \quad \xi = \xi', \quad \eta,$$

we obtain the multiplication table

$$b^2 = 1, \quad b\xi = b\eta = 0, \quad \xi\eta = b,$$

from which it follows that this algebra is isomorphic to the matrix algebra $A_{1,1}$ defined on a space with one even and one odd variable.

It is easy to verify that among those remaining the algebras $\mathscr{D}_{t'}$ and $\mathscr{D}_{t''}$ whose parameters are connected by the relation

$$t' + t'' + 2t't'' = 0 \tag{4.5}$$

are isomorphic. To prove this we pass to the basis

$$1, \quad b = 1 - a, \quad \xi = \frac{1}{1+t'}\xi', \quad \eta.$$

In this basis the multiplication has the form (4.3), if we put $t = t'/(1+t')$. It is obvious that if in the table (4.3) the parameters t and $\tilde{t}$ differ in sign, then the algebras are isomorphic (it suffices to put $\tilde{b} = -b$). In table (4.4) this condition agrees with (4.5).

On the other hand, the vector $b' \neq 1$ satisfying $(b')^2 = 1$ is uniquely determined up to sign: $b' = \pm b$. Therefore if we append to table (4.3) the condition $t \geq 0$, we obtain a list of the nonisomorphic algebras of the series $\mathscr{D}_t$.

The above observations can be summarized in the following theorem.

THEOREM 4.1 (correction of the classification of Jordan superalgebras). *The list of Jordan superalgebras given in* [1], *together with the series of Jordan superalgebras* H_n $(n \geq 3)$ *and the corrections in items* A *and* B, *provides a complete description, to within isomorphism, of the simple finite-dimensional Jordan superalgebras over an algebraically closed field of characteristic zero.*

REMARK. The algebra H_2 is isomorphic to $A_{1,1}$.

Let us consider the problem of classifying Poisson algebras. We first make the following observation. If a Poisson algebra U with operations

$$xy, \qquad \{x, y\} \tag{4.6}$$

is simple, then so is the Poisson algebra $\widetilde{U}$ with operations

$$xy, \quad \lambda\{x, y\}, \quad \lambda \neq 0, \tag{4.7}$$

which is not isomorphic to U.

The presence of the parameter λ corresponds to the fact that in the definition of an associative Jordan algebra of the form $H(\sigma, S)$ the linear transformation S is only defined to within a factor.

In Theorem 4.2 and other theorems on the classification of Poisson algebras, the algebras are considered to within multiplication of one of the operations by $\lambda \neq 0$.

THEOREM 4.2. *A simple finite-dimensional Poisson algebra with nonzero operations xy and $\{x, y\}$ over an algebraically closed field of characteristic zero is isomorphic, to within multiplication of one of the operations by $\lambda \neq 0$, to a Poisson algebra H_n $(n \geq 2)$.*

PROOF. According to Theorem 3.4, we must isolate from the simple Jordan algebras the associative Jordan algebras of the form $H(\sigma, S)$. The first condition, the existence of an associative subalgebra of "half" the dimension, leaves only the series of algebras H_n and the algebras $A_{1,1}$, $\mathscr{D}_t$ of small dimension. We will regard the algebra $A_{1,1} \simeq H_2$ as an algebra in the series H_n.

Regarding the algebras $\mathscr{D}_t$, it is easy to see that the second condition, $S(xy) = S(x)y$, is not satisfied for the possible associative subalgebras $\{1, b\}$ and $\{1 \pm b, \xi\}$ (see (4.3)).

For the algebras H_n, the complement to the associative subalgebra defining the involutive automorphism σ is uniquely determined. Also uniquely determined is the transformation S, since $S(x) = S(1) * x$ and the element $S(1)$ must satisfy $S(1) * S(x) = 0$, $\forall x$; such an element is unique in the complementary plane.

We can now write down the operations xy and $\{x, y\}$ in accordance with (3.7). We obtain the Poisson algebra H_{n-1}. The theorem is proved.

In the next section we will prove a generalization of Theorem 4.2.

§5. Poisson-Jordan algebras

The proofs of the theorems in §3 and §4 are almost independent of the property of associativity of Poisson algebras. There naturally arises the thought of formulating a more general concept including the ordinary Poisson algebras as a special case. This is done in this section.

We first consider another important special case of this more general concept.

DEFINITION 5.1. By a Poisson algebra of the second type we mean a linear space with two bilinear operations, a commutative operation $(a, b) \mapsto ab$ and a skew-symmetric operation $(a, b) \to \{a, b\}$, jointly satisfying the Leibniz identity

$$\{ab, c\} = a\{b, c\} + (-1)^{ab} b\{a, c\} \tag{5.1}$$

and the identity

$$(a, b, c) = (-1)^{ab} \lambda\{b, \{a, c\}\}, \tag{5.2}$$

where (a, b, c) denotes the associator, i.e.,

$$(a, b, c) = (ab)c - a(bc), \tag{5.3}$$

and $\lambda \neq 0$ is a fixed number.

An example of a Poisson algebra of the second type is the space of square matrices of order n with the operations

$$A \cdot B = AB + BA \quad \text{and} \quad \{A, B\} = AB - BA.$$

Another important example arises in quantum mechanics. It is the space of Hermitian operations with the operations

$$A\cdot B = AB + BA \quad \text{and} \quad \{A\,,\,B\} = \frac{n}{i}(AB - BA)\,.$$

It follows from identity (5.2) that the skew-symmetric operation is Lie. To prove this we need only express all three terms of the Jacobi identity in terms of associators in accordance with (5.2).

It is also easy to deduce from identities (5.1) and (5.2) that the commutative operation is Jordan, but this is a consequence of the following more general assertion.

PROPOSITION 5.1. *Suppose the field over which a Poisson algebra of the second type is defined contains* $\sqrt{\lambda}$. *Then there exists on the space of the Poisson algebra of the second type an associative operation* $a\times b$ *such that*

$$a\cdot b = \frac{1}{2}(a\times b + b\times a)\,, \qquad \{a\,,\,b\} = \frac{1}{2\sqrt{\lambda}}(a\times b - b\times a)\,. \tag{5.4}$$

Conversely, the space of an associative algebra with two operations defined by (5.4) *is a Poisson algebra of the second type.*

PROOF. The operation $a\times b$ can obviously be uniquely reconstructed from the formulas (5.4):

$$a\times b = ab + \sqrt{\lambda}\{a\,,\,b\}\,.$$

We will show that the associator of this operation is zero:

$$\begin{aligned}
(a\times b)\times c - a\times(b\times c)&\\
&= (ab + \sqrt{\lambda}\{a\,,\,b\})c + \sqrt{\lambda}\{\underline{\underline{ab}} + \sqrt{\lambda}\underline{\{a\,,\,b\}}\,,\,c\}\\
&\quad - a(bc + \sqrt{\lambda}\{b\,,\,c\} - \sqrt{\lambda}\{a\,,\,\underline{\underline{bc}} + \sqrt{\lambda}\underline{\{b\,,\,c\}}\}\\
&= [(ab)c - a(bc) - (-1)^{ab}\lambda\underline{\{b\,,\,\{a\,,\,c\}\}}] + \sqrt{\lambda}[\{a\,,\,b\}c - a\{b\,,\,c\}\\
&\qquad\qquad - (-1)^{bc}\underline{\underline{\{ac\,,\,b\}}}] = 0.
\end{aligned}$$

Here the terms underlined with a single line are transformed by means of the Jacobi identity, and the terms underlined with two lines by means of the consequence (1.6) of the Leibniz identity. Also, the first square bracket is equal to zero in view of (5.2), and the second in view of (5.1).

The proof of the converse can be obtained by direct substitution of the formulas (5.4) into (5.2) and (5.1) and by use of the associativity of the operation $a\times b$. We omit these calculations.

The following definition removes the condition that the commutative operation be associative.

DEFINITION 5.2. By a generalized Poisson algebra we mean a linear space with two bilinear operations, a commutative operation $(a\,,\,b)\mapsto ab$ and a Lie operation $(a\,,\,b)\mapsto\{a\,,\,b\}$, jointly satisfying the Leibniz rule

$$\{ab\,,\,c\} = a\{b\,,\,c\} + (-1)^{ab}b\{a\,,\,c\}\,. \tag{5.5}$$

In the next section we will return to this definition but we now add two conditions.

Definition 5.3. We call a generalized Poisson algebra a Poisson-Jordan algebra if it satisfies the following two additional conditions:

1) $x \cdot y$ is a Jordan operation,

2)

$$\{(x, y, z), u\} = (-1)^{y(z+u)}(\{x, z\}, u, y), \tag{5.6}$$

where $(x, y, z) = (xy)z - x(yz)$ is the associator of a commutative operation.

Remark. From (5.6) it follows that the Jordan identity (1.12) holds, up to a value from the center Z of the skew symmetric operation $\{x, y\}$. However, it is quite impossible to omit condition 1).

It is easy to verify that both the ordinary Poisson brackets and the Poisson brackets of the second type satisfy this definition. In the first case it follows from associativity that the left and right sides of (5.6) are separately equal to zero. In the second case, substituting the expression (5.2) for the associator into the left and right sides we arrive at the identity.

We show that all the theorems of §3 formulated for Poisson brackets also remain valid for Poisson-Jordan brackets.

Definition 5.4. Let U be a Poisson-Jordan algebra. The notation $\mathfrak{J}(U)$ refers to the algebra arising from the space $U + U^S$ with the operation given by the formulas

$$\begin{aligned} a * b = ab, \qquad a^S * b = (-1)^a a * b^S = (ab)^S, \\ a^S * b^S = (-1)^a\{a, b\}, \quad \forall a, b \in u, \ a^S, b^S \in u^S. \end{aligned} \tag{5.7}$$

Theorem 5.1. *$\mathfrak{J}(U)$ is a Jordan (super)algebra.*

The proof is almost a verbatim repetition of the proof of Theorem 3.1. Associativity is used in two places. First, it is used in the verification of the Jordan identity when $a, b, c \in U$. Its validity follows from the fact that the commutative operation on U is Jordan and from the formula $ab^S = (-1)^a(ab)^S$. Second, associativity is used in the verification of the Jordan identity when $a \in U$, $b = x^S$, $c = y^S \in U^S$. In this case, instead of relation (3.6) we have

$$Q_{x(ya)} - (-1)^{xy}Q_{y(xa)} = (-1)^{a(x+y)}[P_a, P_{\{x, y\}}], \tag{5.8}$$

which follows from (5.6). Since associativity has not been used anywhere, Theorem 5.1 has been proved.

To formulate the following theorems we will weaken Definition 3.2 by removing associativity.

Definition 5.5. We will say that a Jordan algebra A is a Jordan algebra of the form $H(\sigma, S)$ if A satisfies conditions 2) and 3) of Definition 3.2.

Definition 5.6. Suppose A is a Jordan algebra of the form $H(\sigma, S)$. By the algebra $P(A)$ we mean the algebraic system on the space A_1 with the two operations

$$xy = x * y, \qquad \{x, y\} = (-1)^{x}(S(x) * S(y)). \tag{5.9}$$

THEOREM 5.2. *Suppose A is a Jordan algebra of the form $H(\sigma, S)$. Then the algebra $P(A)$ is a Poisson-Jordan algebra.*

The proof repeats verbatim that of Theorem 3.2. However, it is now necessary to verify the additional identity (5.6). If we substitute $a \in A_1$, $b = u^S$, $c = v^S \in A_{-1}$ in (1.12) and take $x^S \in A_{-1}$ as the argument, we obtain (after an application of the Leibniz identity)

$$Q_{u(va)} - (-1)^{uv} Q_{v(u,a)} = (-1)^{a(u+v)}[P_a, P_{\{u,v\}}], \tag{5.10}$$

which is easily seen to agree with (5.6). The theorem is proved.

From Theorems 5.1 and 5.2 we obtain

THEOREM 5.3. *The mappings P and $\mathfrak{J}$ are mutually inverse and establish a one-to-one correspondence between the Poisson-Jordan algebras and the Jordan algebras of the form $H(\sigma, S)$. This correspondence is functorial.*

The proof is analogous to that of Theorem 5.3.

THEOREM 5.4. *Suppose U is a Poisson-Jordan algebra with nonzero operations xy and $\{x, y\}$. Then the Jordan algebra $\mathfrak{J}(U)$ is simple if and only if U is simple.*

The proof agrees completely with the proofs of Lemma 3.1 and Theorem 3.4, since nowhere in those proofs is the associativity of the operation xy used.

Let $A_{m,n}$ be the associative superalgebra of matrices acting on a space with m odd and n even variables. We denote by the same symbol $A_{m,n}$ the Poisson algebra of the second type with operations

$$x \cdot y = xy + (-1)^{xy} yx, \qquad \{x, y\} = xy - (-1)^{xy} yx. \tag{5.11}$$

Analogously, let Q_n denote the Poisson algebra of the second type defined by the formulas (5.11) on the space of matrices of the form $\left(\begin{smallmatrix} a & b \\ b & a \end{smallmatrix}\right)$, where a is an even and b an odd square matrix of order n.

THEOREM 5.5. *A simple finite-dimensional Poisson-Jordan algebra over an algebraically closed field of characteristic zero is isomorphic to one of the brackets of the three series $A_{m,n}$, Q_n, H_n, to within multiplication of one of the operations by $\lambda \neq 0$.*

PROOF. According to Theorem 5.4, we must find involutive automorphisms of simple Jordan algebras satisfying the additional condition that there exists an odd one-to-one mapping S of the subalgebra A_1 onto the complementary plane A_{-1} such that $S(x \cdot y) = S(x) \cdot y$. With this in mind, we pass from the simple Jordan algebra A to the simple Lie superalgebra $\mathfrak{L}(A)$ by means of the construction given in [1], [5], [6]. In this regard, to an involutive automorphism of A there corresponds an involutive automorphism of $\mathfrak{L}(A)$, and the above-mentioned additional condition becomes the following: the representation of a stationary subalgebra of an involutive automorphism of $\mathfrak{L}(A)$ on the complementary plane (with parity of vectors changed into the opposite and to within an annihilating subspace) must be equivalent to the adjoint representation.

All involutive automorphisms of simple Lie superalgebras are listed in [7], [8]. An examination of these automorphisms reveals that the representation on the complementary plane is equivalent to the adjoint for only three types of involutive automorphisms of the Lie superalgebras Q_n, $A_{n,n}$, H_n. The corresponding involutive automorphisms of the Jordan algebras are the following: for the Jordan algebra Q_n, an automorphism with stationary subalgebra of matrices of the form

$$\left(\begin{array}{cc|cc} B_1 & 0 & 0 & C_1 \\ \hline 0 & B_2 & C_2 & 0 \\ \hline 0 & C_1 & B_1 & 0 \\ \hline C_2 & 0 & 0 & B_2 \end{array}\right);$$

for the Jordan algebra $A_{n,n}$, an automorphism with stationary subalgebra of matrices, and for the Jordan algebra H_n, the standard involutive automorphism corresponding to the construction of the Jordan algebra H_n in §4.

Using (5.9), we can now write down the operations in the corresponding Poisson-Jordan algebras. There appear the three series of Poisson-Jordan algebras indicated in the theorem:

$$A_{m,n} = P(Q_{m+n}), \quad Q_n = P(A_{n,n}), \quad H_n = P(H_{n+1}).$$

The theorem is proved.

In the next section we will prove a more general result, but under a stronger restriction on the simplicity of the bracket.

§6. Generalized Poisson algebras

In this section we will construct from a generalized Poisson algebra not a Jordan superalgebra, but a Lie superalgebra. In principle, the constructions and proofs are the same as in §3 and §5, but the roles of the commutative and skew-symmetric operations of the Poisson algebra are interchanged.

However, because of the asymmetry of these operations in the definition of a Poisson algebra the assertions appear to be somewhat different.

Suppose U is the space of a generalized Poisson algebra. Consider the direct sum $U \dot{+} U^S$, where U^S is a duplicate of U with the opposite parity of elements. We define on this space a commutation operation [,] by the formulas

$$\begin{gathered}[x, y] = \{x, y\}, \qquad [x^S, y] = (-1)^x [x, y^S] = \{x, y\}^S, \\ [x^S, y^S] = (-1)^x x \cdot y, \quad \forall x, y \in U, \ x^S, y^S \in U^S.\end{gathered} \tag{6.1}$$

DEFINITION 6.1. By the algebra $\mathfrak{L}(U)$, where U is a generalized Poisson algebra, we mean the space $U \dot{+} U^S$ with the operation (6.1).

THEOREM 6.1. *The algebra $\mathfrak{L}(U)$ is a Lie (super)algebra.*

PROOF. The anticommutativity of the operation [,] is a consequence of the following simple calculations:

$$[a^S, b] = \{a, b\}^S = -(-1)^{ab}\{b, a\}^S = (-1)^{ab+b}[b, a^S] = -(-1)^{a^S b}[b, a^S],$$

$$[a^S, b^S] = (-1)^a a \cdot b = (-1)^{a+ab} ba = -(-1)^{a^S b^S + b} b \cdot a = -(-1)^{a^S b^S}[b^S, a^S].$$

Recall that we have denoted by P_a and Q_a the linear "shift" operators on the space U:

$$P_a(x) = a \cdot x, \qquad Q_a(x) = \{a, x\}.$$

We can verify the following relations:

$$\begin{gathered}[Q_a, Q_b] = Q_{\{a,b\}}, \qquad [P_a, Q_b] = P_{\{a,b\}}, \\ P_a Q_b + (-1)^{ab} P_b Q_a = Q_{ab}, \qquad Q_a P_b + (-1)^{ab} Q_b P_a = Q_{ab}.\end{gathered} \tag{6.2}$$

The first is a consequence of the Jacobi identity, the second and third follow from the Leibniz identity, and the fourth is a consequence of (1.6).

It follows from the definition of the algebra $\mathfrak{L}(U)$ that the left shifts $L_a(x) = [a, x]$ in this algebra can be represented by the "second-order" matrices

$$L_a = \left(\begin{array}{c|c} Q_a & 0 \\ \hline 0 & (-1)^a Q_a \end{array}\right), \quad L_{a^S} = \left(\begin{array}{c|c} 0 & (-1)^a P_a \\ \hline Q_a & 0 \end{array}\right), \quad a \in U, \ a^S \in U^S.$$

Considering the commutators of left shifts and using (6.2), we obtain

$$[L_a, L_b] = \left(\begin{array}{c|c} [Q_a, Q_b] & 0 \\ \hline & (-1)^{a+b}[Q_a, Q_b] \end{array}\right) = \left(\begin{array}{c|c} Q_{\{a,b\}} & 0 \\ \hline 0 & (-1)^{a+b} Q_{a+b} \end{array}\right)$$

$$[L_a, L_{b^S}] = \left(\begin{array}{c|c} 0 & (-1)^b Q_a P_b - (-1)^{ab^S+b+a} P_b Q_a \\ \hline (-1)^a Q_a Q_b - (-1)^{ab^S} Q_b Q_a & 0 \end{array}\right) = \left(\begin{array}{c|c} 0 & (-1)^a P_{\{a,b\}} \\ \hline (-1)^a Q_{\{a,b\}} & 0 \end{array}\right) = (-1)^a L_{\{a,b\}^S}$$

$$[L_{a^S}, L_{b^S}] = \left(\begin{array}{c|c} (-1)^a P_a Q_b - (1-)^{a^S b^S+b} P_b Q_a & 0 \\ \hline 0 & (-1)^b Q_b P_b - (-1)^{a^S b^S+a} Q_b P_a \end{array}\right) = \left(\begin{array}{c|c} (-1)^a Q_{ab} & 0 \\ \hline 0 & (-1)^b Q_{ab} \end{array}\right) = (-1)^a L_{ab}. \tag{6.3}$$

It follows from these three equalities that for any $x, y \in U \dotplus U^S$ we have

$$[L_x, L_y] = L_{[x,y]}, \tag{6.4}$$

i.e., the algebra $\mathfrak{L}(U)$ satisfies the Jacobi identity. The theorem is proved.

To invert the correspondence $U \mapsto \mathfrak{L}(U)$ we make the following

DEFINITION 6.2. Suppose σ is an involutive automorphism of a Lie superalgebra A such that for a decomposition $A = A_1 \dotplus A_{-1}$ of A into σ-invariant subspaces there exists a one-to-one odd linear mapping $S\colon A_1 \to A_{-1}$.

The Lie superalgebra A will be called a Lie algebra of the form $H(\sigma, S)$ if

$$S([x, y]) = [S(x), y]. \tag{6.5}$$

DEFINITION 6.3. Suppose A is a Lie superalgebra of the form $H(\sigma, S)$. By the algebra $P(A)$ we mean the algebraic system on the subspace A_1 with the two operations

$$\{x, y\} = [x, y], \qquad x \cdot y = (-1)^x [x^S, y^S], \quad x, y \in A_1. \tag{6.6}$$

THEOREM 6.2. *The algebra $P(A)$ is a generalized Poisson algebra.*

PROOF. The skew-symmetry of the operation $\{x, y\}$ and the Jacobi identity follow from the fact that A_1 is a subalgebra of the Lie superalgebra A. Commutativity of the operation xy can be verified in a simple way:

$$x \cdot y = (-1)^x [x^S, y^S] = -(1)^{(x+1)(y+1)+x} [y^S, x^S] = (-1)^{xy} y \cdot x.$$

To prove the Leibniz identity we consider the Jacobi identity in A for $x^S, y^S \in A_1$ and $a \in A_1$:

$$[[x^S, y^S], a] = [x^S, [y^S, a]] - (-1)^{(x+1)(y+1)}[y^S, [x^S, a]].$$

Writing this relation in terms of $P(A)$, we obtain

$$(-1)^x\{x \cdot y, a\} = (-1)^x x \cdot \{y, a\} + (-1)^{x+xy} y \cdot \{x, a\},$$

i.e., we obtain the Leibniz identity. The theorem is proved.

From Theorems 6.1 and 6.2 there follows

THEOREM 6.3. *The mappings P and $\mathfrak{L}$ are mutually inverse and establish a one-to-one correspondence between the generalized Poisson algebras and the Lie superalgebras of the form $H(\sigma, S)$. This correspondence is functorial.*

PROOF. An elementary comparison of formulas (6.1) and (6.6) reveals that for a generalized Poisson algebra U and a Lie superalgebra A of the form $H(\sigma, S)$ we have

$$\mathfrak{L}(P(A)) = A, \qquad P(\mathfrak{L}(U)) = U. \tag{6.7}$$

By an S-ideal of a Lie superalgebra $A \in H(\sigma, S)$ we mean an ideal of the form

$$\mathscr{D}^* = \mathscr{D} \dotplus S(\mathscr{D}),$$

where $\mathscr{D}$ is an ideal of the subalgebra A_1. The factor algebra $A/\mathscr{D}^*$ is an algebra of the form $H(\sigma, S)$; the homomorphism $A \to A/\mathscr{D}^*$ will be called an S-homomorphism.

Consider two categories: the category $\widetilde{P}$ of generalized Poisson algebras with homomorphisms as morphisms, and the category of Lie algebras of the form $H(\sigma, S)$ with S-homomorphisms as morphisms. Then the following two diagrams are commutative:

$$\begin{array}{ccc} U & \xrightarrow{\mathfrak{L}} & \mathfrak{L}(U) \\ \downarrow{\scriptstyle\varphi} & & \downarrow{\scriptstyle\varphi'} \\ U' & \xrightarrow{\mathfrak{L}} & \mathfrak{L}(U') \end{array}, \qquad \begin{array}{ccc} A & \xrightarrow{P} & P(A) \\ \downarrow{\scriptstyle\psi} & & \downarrow{\scriptstyle\psi'} \\ A' & \xrightarrow{P} & P(A') \end{array}, \tag{6.8}$$

since an ideal $\mathscr{D}$ of a generalized Poisson algebra $P(A)$ is uniquely determined by an S-ideal of A, and an ideal $\mathscr{D}^*$ of $\mathfrak{L}(U)$ can be uniquely constructed from an ideal of the generalized algebra U. The theorem is proved.

It follows from Theorem 6.3 that if a generalized Poisson algebra U is simple, then the Lie algebra $\mathfrak{L}(U)$ contains no S-ideals. We will prove a more general result.

DEFINITION 6.4. By a weak ideal of a generalized Poisson algebra we mean a subspace $\mathscr{D}$ such that

$$\{\mathscr{D}, x\} \subset \mathscr{D}, \qquad \{\mathscr{D} \cdot x, y\} \subset \mathscr{D}, \quad \forall x, y. \tag{6.9}$$

REMARK. It is obvious that an ordinary ideal of a generalized Poisson algebra, i.e., a joint ideal of both operations, is a weak ideal.

DEFINITION 6.5. Two generalized Poisson algebras U_1, U_2 will be called equivalent if there exists a one-to-one mapping $\varphi: U_1 \to U_2$ such that

$$\begin{gathered} \{\varphi(x), \varphi(y)\}_2 = \varphi(\{x, y\}_1), \\ (\varphi(x) \cdot \varphi(y))_2 \equiv \varphi((x \cdot y)_1) \mod Z, \end{gathered} \tag{6.10}$$

where Z is the center of the skew-symmetric operations $\{x, y\}_2$ of U_2.

In other words, two algebras are equivalent if there exists a mapping φ effecting 1) an isomorphism of the skew-symmetric operations, and 2) an isomorphism of the symmetric operations to within a summand in the center of the skew-symmetric operation.

PROPOSITION 6.1. *The mapping $U \to U/\mathscr{D}$, where $\mathscr{D}$ is a weak ideal of the generalized Poisson algebra U, defines on the factor space $U/\mathscr{D}$ the structure of a generalized Poisson algebra, defined to within equivalence.*

PROOF. Let $\mathscr{D}_1$ denote a maximal subspace such that

$$\{\mathscr{D}_1, U\} \subset U. \tag{6.11}$$

Then it follows from the definition of a weak ideal that

$$U \cdot \mathscr{D} \subset \mathscr{D}_1. \tag{6.12}$$

Relation (6.11) means that $\mathscr{D}_1/\mathscr{D}$ is the center of the skew-symmetric operation defined on the factor space $U/\mathscr{D}$. It therefore follows from (6.12) that the symmetric operation on $U/\mathscr{D}$ is defined to within terms in Z. The proposition is proved.

Theorem 6.5 below can be regarded as a classification of the simple objects of the category of generalized Poisson algebras, viewed to within equivalence.

THEOREM 6.4. *Suppose U is a generalized Poisson algebra with nonzero operations $x \cdot y$ and $\{x, y\}$. Then the Lie superalgebra $\mathfrak{L}(U)$ is a central extension of a simple Lie superalgebra if and only if U contains no weak ideals.*

The center Z of $\mathfrak{L}(U)$ is contained in the subalgebra A_1 and is equal to the center of the skew-symmetric operation of U.

PROOF. The "only if" part follows from the definition of multiplication in $\mathfrak{L}(U)$. If $\mathscr{D}$ is a weak ideal of U and $\widehat{\mathscr{D}} = U \cdot \mathscr{D}$, then the subspace $\widehat{\mathscr{D}} \dotplus \mathscr{D}^S$ is an ideal of $\mathfrak{L}(U)$.

To prove the converse, consider any ideal $\mathscr{D}$ of $\mathfrak{L}(U)$ and denote by $\mathscr{D}_1$ and $\mathscr{D}_2^S$ its projections on the subspaces U and U^S, respectively. Commuting $\mathscr{D}$ with arbitrary elements $a \in U$ and $a^S \in U^S$, we obtain the relations

$$\{a, \mathscr{D}_1\} \subset \mathscr{D}_1, \quad \{a, \mathscr{D}_2\} \subset \mathscr{D}_2, \quad \{a, \mathscr{D}_1\} \subset \mathscr{D}_2, \quad a\mathscr{D}_2 \subset \mathscr{D}_1. \tag{6.13}$$

It follows that $\mathscr{D}_1 \cap \mathscr{D}_2$ is a weak ideal of U. There are two possibilities: 1) $\mathscr{D}_1 \cap \mathscr{D}_2 = 0$; 2) $\mathscr{D}_1 \cap \mathscr{D}_2 = U$.

Consider the first, $\mathscr{D}_1 \cap \mathscr{D}_2 = 0$. It follows from (6.13) that $\mathscr{D}_1$ is contained in the center of the Lie algebra with operation $\{x, y\}$. If $\mathscr{D}_2 \neq 0$, then, in view of (6.13), $\mathscr{D}_2$ is a weak ideal of U. The case $\mathscr{D}_1 = 0$, $\mathscr{D}_2 = U$ is impossible, since the multiplication xy is nonzero. Consequently, $\mathscr{D}_2 = 0$ and the ideal $\mathscr{D} = \mathscr{D}_1$ is contained in the center Z of the algebra $\mathfrak{L}(U)$ mentioned in the theorem.

Consider the second possibility, $\mathscr{D}_1 \cap \mathscr{D}_2 = U$, and distinguish in the ideal $\mathscr{D}$ a maximal subspace T containing the pairs $t \in U$ and $t^S \in U^S$. The intersection T_1 of T and U is a weak ideal of the Poisson algebra U. Indeed, from

$$[x, T_1] = \{x, T_1\} \subset U, \qquad [x^S, T_1] = \{x, T_1\}^S \subset U^S$$

we obtain $\{x, T_1\} \subset T_1$, and from

$$[y, [x^S, T_1]] = \{y, x \cdot T_1\} \subset U,$$
$$[y^S, [x^S, T_1^S]] = \{y, x \cdot T_1\}^S \subset U^S, \quad \forall x, y,$$

we obtain $\{y, x \cdot T_1\} \subset T_1$, $\forall x, y$, i.e., T_1 is a weak ideal, hence $T_1 = 0$.

Since $T_1 = 0$ and $\mathscr{D}_1 \cap \mathscr{D}_2 = U$, it follows that the elements of $\mathscr{D}$ have the form $a + (f(a))^S$, where $a \in U$, $(f(a))^S \in U^S$, and f is a one-to-one odd linear mapping of U into itself. Multiplying the elements of $\mathscr{D}$ on the right by $b \in \mathscr{U}$, we obtain the relation

$$f(\{a, b\}) = \{f(a), b\}. \tag{6.14}$$

Multiplying the elements of $\mathscr{D}$ on the right by $b^S \in U^S$, we obtain

$$f(f(a) \cdot b) = -\{a, b\}. \tag{6.15}$$

Using (6.14) and (6.15) and transforming $f^2(a \cdot b)$, we see that

$$f^2(a \cdot b) = f(f(f(f^{-1}(a)) \cdot b)) = -f(\{f^{-1}(a), b\}) = -\{a, b\},$$

i.e.,

$$f^2(a \cdot b) = -\{a, b\}.$$

This last equality is impossible, since the symmetric bilinear operation $f^2(a \cdot b)$ cannot be equal to the nonzero skew-symmetric bilinear operation $-\{a, b\}$. The theorem is proved.

It is easy to verify that the subspace $U' = \{U, U\}$ is a weak ideal of the Poisson algebra U. Consequently, for a Poisson algebra without weak ideals we have $U' = U$. For this reason, the simple Poisson-Jordan algebras $A_{m,n}$, Q_n, H_n contain a weak ideal.

Let $\widetilde{A}_{m,n}$ denote the Poisson algebra defined on the space of all matrices with supertrace zero, acting on a space with m even and n odd variables with the operations

$$\{A, B\} = AB - BA,$$
$$A * B = AB + BA - \frac{1}{m+n}\operatorname{str}(AB + BA).$$

THEOREM 6.5. *A generalized Poisson algebra U without weak ideals is equivalent, to within multiplication of one of the operations by a scalar factor $\lambda \neq 0$, to a generalized Poisson algebra $\widetilde{A}_{m,n}$.*

PROOF. The Lie superalgebra $\mathfrak{L}(U)$ is of the form $H(\sigma, S)$ and, by Theorem 6.4, is a central extension of a simple Lie superalgebra. We should therefore consider involutive automorphisms of simple Lie superalgebras and select from among them those for which the representation of the stationary subalgebra A_1' on the complementary plane A_{-1} agrees, to within the action on the annihilating subspace Z_{-1}, with the adjoint representation of A_1'. The mapping $A_{-1} \to A_1'$ proving the "equivalence" of the representations must be odd.

The involutive automorphisms of simple superalgebras were found in [7], [8]. An examination of these automorphisms reveals that the above restrictions are satisfied only by the involutive automorphism of the Lie superalgebra Q_{n+m} with stationary subalgebra of matrices of the form

$$\left(\begin{array}{cc|cc} B_1 & 0 & 0 & C_1 \\ 0 & B_2 & C_2 & 0 \\ \hline 0 & C_1 & B_1 & 0 \\ C_2 & 0 & 0 & B_2 \end{array}\right), \qquad \operatorname{tr} B_1 + \operatorname{tr} B_2 = 0. \tag{6.16}$$

Note that for the Lie superalgebras $A_{n,n}$ and H_n there exist involutive automorphisms such that the representation of a stationary subalgebra is "almost" equivalent to the adjoint representation. However, in these two cases the representation space contains only a subspace of codimension 1 satisfying the necessary conditions.

The Lie superalgebra Q_{n+m} with stationary subalgebra (6.16) is of the form $H(\sigma, S)$, and its corresponding generalized Poisson algebra is $\widetilde{A}_{m,n} = P(Q_{n+m})$.

A central extension of the Lie superalgebra Q_{n+m} to a new Lie superalgebra of the form $H(\sigma, S)$ is impossible in this case, since the center must be contained in a stationary subalgebra.

The theorem is proved.

BIBLIOGRAPHY

1. V. G. Kac, *Classification of simple Z-graded Lie superalgebras and simple Jordan superalgebras*, Comm. Algebra **5** (1977), 1375–1400.

2. I. L. Kantor, *Some generalizations of Jordan algebras*, Trudy Sem. Vektor. Tenzor. Anal. **17** (1974), 255–313. (Russian)

3. I. L. Kantor and D. B. Persits, *On Jordan and Lie operations connected with a simple Lie superalgebra*, in: Applied Problems of Differential Geometry, Moskov. Oblast. Ped. Inst., Moscow, 1982, pp. 58–68, Manuscript No. 5570–83, deposited at VINITI, 1983. (Russian)

4. L. Hogben, *Erratum*, Comm. Algebra **10** (1983), 1155–1161.

5. I. L. Kantor, *Classification of irreducible transitive differential groups*, Dokl. Akad. Nauk SSSR **158** (1964), 1271–1274; English transl. in Soviet Math. Dokl. **5** (1964).

6. ——, *Transitive differential groups and invariant connections in homogeneous spaces*, Trudy Sem. Vektor. Tenzor. Anal. **13** (1966), 310–398. (Russian)

7. V. V. Serganova, *Classification of simple real Lie superalgebras and symmetric superspaces*, Funktsional Anal. i Prilozhen. **17** (1983), no. 3, 46–54; English transl. in Functional Anal. Appl. **17** (1983), no. 3.

8. ——, *Automorphisms and real forms of simple Lie superalgebras*, Candidate's Dissertation Phys.-Math. Sciences, Leningrad State University, Leningrad, 1987. (Russian)

Translated by G. A. KANDALL

Amer. Math. Soc. Transl.
(2) Vol. **151**, 1992

New Methods and Trends in Geometric Function Theory

UDC 517.54:514.763.42:515.171.5

S. L. KRUSHKAL′

§1

One of the features of the modern development of geometric function theory, as in other divisions of analysis, is the close linkage of its methods with the ideas and approaches from other, seemingly separate, areas of mathematics. This theme is very extensive. I will dwell here just on the applications of geometric methods of multidimensional complex analysis, complex differential geometry, and the theory of Teichmüller spaces to the solution of a series of well-known problems connected with conformal and quasiconformal mappings that have stood for a rather long time. In this way a whole direction in geometric function theory is developed at once. The traditional methods of function theory give nothing here; progress has been obtained thanks to completely new methods.

For example, it has been possible to find an explicit form for the extremals of many variational problems, to solve the question of the relation between necessary and sufficient conditions for quasiconformal extension of analytic functions, etc. At the basis of these results lies, as it has turned out, properties of hyperbolic metrics on Teichmüller spaces, first and foremost the Carathéodory metric. A partial account may be found in [7] and [8].

§2

We consider a different circle of problems in more detail. They are connected with problems of two-dimensional potential theory, the theory of Fredholm eigenvalues, and the theory of approximation of functions. The question is the extension of conformal mappings.

1991 *Mathematics Subject Classification*. Primary 32G15; Secondary 58B20, 30C60.

In various questions of complex analysis and its applications one has to extend conformal mappings of plane regions continuously to the whole Riemann sphere $\widehat{\mathbb{C}} = \mathbb{C} \cup \{\infty\}$. In addition, it is often desirable to have quasiconformal extensions, and then it is important to know an estimate (best, if possible) of the dilatations of these extensions. Such extensions reduce to quasiconformal reflection with respect to curves that are images of a circle, and again the problem arises of an estimate of the least of the possible reflection coefficients for the given curve.

For example, the construction of a conformal mapping of a region with a smooth boundary reduces to the solution of an integral equation with double layer potential. Its eigenvalues are just Fredholm. They are encountered in various problems. In some problems the least positive eigenvalue λ_1 plays a critical role; it is estimated exactly by the reflection coefficient (see, for example, [15], [3]), moreover not just for smooth curves.

In the case of analytic curves, which is very important and will be studied here, the solution of the problems under consideration reduces to the investigation of quasiconformal extensions of interpolating families for a univalent analytic function, contracting it to the identity transformation.

Without loss of generality, it is possible to consider normalized univalent functions; in this connection it will be technically more convenient to take the hydrodynamic normalization.

Let Σ be the class of univalent functions meromorphic in $\Delta^* = \{z \in \widehat{\mathbb{C}}: |z| > 1\}$ with expansion

$$f(z) = z + \sum_{n=0}^{\infty} a_n z^n,$$

Σ_k the subclass of functions from Σ admitting a k'-quasiconformal extension to the unit disk $\Delta = \{z: |z| < 1\}$ with $k' \leq k$ $(0 \leq k < 1)$.

We consider the interpolating family

$$f_r(z) = zf\left(\frac{z}{r}\right), \qquad 0 \leq r < 1 \quad (f_0(z) \equiv z).$$

Each function f_r, together with f, automatically belongs to Σ_k, but it turns out that it is possible to take the extension of f_r to $\overline{\Delta}$ with a substantially smaller dilatation.

The problems of estimating the dilatation of extensions and reflections have been considered for a long time; in the strongest statement they are formulated in the work of Kühnau (see, for example, [12], [13]). The formulation of the Kühnau problems is the following:

PROBLEM 1. If $f \in \Sigma_k$, does f_r belong to $\sum_{kr^2}$? Specifically is it true that

$$f \in \Sigma \quad \Rightarrow \quad f_r \in \Sigma_{r^2}? \tag{1}$$

(This case is especially important for applications.)

PROBLEM 2. Let a curve L be the image of the unit circle $\{|z| = 1\}$ under a conformal mapping of the annulus $\{r < |z| < R\}$. Does it admit a

q_L-quasiconformal reflection with

$$q_L \leq \frac{1+(rR)^2}{r^2+R^2} \qquad (q_L < 1)? \tag{2}$$

Often one has to impose supplementary conditions on the quasiconformal extension $\hat{f}$ of the given function f, for example, to demand that $\hat{f}(0)=0$. We set $\Sigma_k(0) = \{f \in \Sigma_k : \hat{f}(0) = 0\}$. For this class the following problem remains, being an equivalent formulation of a problem of Bshouty 6.79 from [2].

PROBLEM 3. Prove or disprove:

$$f \in \Sigma_k(0) \quad \Rightarrow \quad f_r \in \Sigma_{kr}(0). \tag{3}$$

This shows that the presence of a supplementary condition on the extension changes the magnitude of the dilatation $k(f_r) = \|\mu f_r\|_\infty$ in order relative to r (here $\mu_f = f_{\bar{z}}/f_z$ is the Beltrami coefficient of f).

We note that the estimates in (1)–(3) are on the whole not improvable, as one easily convinces oneself by the example of the function

$$f_r(z) = z - 2r + r^2 z^{-1} : \Delta^* \to \widehat{\mathbb{C}}.$$

Previously only very partial results were known for Problems 1 and 3 (Becker [1], Kühnau [12], Hübner [12]). The estimate (2) is suggested by known results on the least nontrivial Fredholm eigenvalue λ_1 for the curve L and the connection of the Fredholm eigenvalues with quasiconformal mappings (concerning this see, for example, [12], [15]).

§3

What are the results obtained here? We consider the functions $f_t(z) = tf(z/t)$ for all complex $t \in \Delta$.

THEOREM 1. *If $f \in \Sigma$, then $f_t \in \Sigma_{|t|^2}$. For $f \in \Sigma_k$ we also have $f_t \in \Sigma_{k|t|}$ $(0 < k < 1)$.*

COROLLARY. *Let f be a conformal mapping of the disk $\{|z| < R\}$, $R > 1$, and $L = f(|z| = 1)$. Then f admits a k-quasiconformal extension in $\overline{\Delta}^*$, and the curve L admits a k-quasiconformal reflection with $k \leq 1/R^2$.*

As has already been said, this estimate is sharp. The original quantitative estimate, obtained in [6], had the form

$$k = (1+k)/(1-k) \leq \nu_0(0, 1/R)^2,$$

where $\nu_0(0, r) = 1 + 2\sum_{n=1}^{\infty}(-1)^n r^{n^2}$ (in particular, for large R it gives just $k \leq 2R^{-1} + 2R^{-2} + \cdots$).

THEOREM 2. *If* $f \in \Sigma_k(0)$, *then* $f_t \in \Sigma_{k|t|}(0)$ *(from which we have also* $f \in \Sigma \Rightarrow f_t \in \Sigma_{|t|}(0)$).

THEOREM 3. *If the curve* L *is the image of the circle* $\{|z| = 1\}$ *under a conformal mapping of the annulus* $\{r < |z| < R\}$, *then it admits a* q_L-*quasiconformal reflection with* q_L *satisfying inequality* (2).

Thus we have a complete solution of Problems 2 and 3, and a solution of Problem 1 in a somewhat weakened form.

§4

The proofs of the above theorems are based on deep properties of Teichmüller spaces, connected with their Finsler and complex structures, and properties of their hyperbolic Kobayashi metric.

Here it is appropriate to introduce the required definitions and results. First of all, the Teichmüller space $T(X)$ of the Riemann surface X is the set of quasiconformal deformations X^μ of the conformal structure of X, modulo homotopy equivalence of the quasiconformal homeomorphisms $X \to X^\mu$. Each quasiconformal homeomorphism W^μ of the surface X satisfies a Beltrami equation $\partial_{\bar z} W = \mu \partial_z W$, whose coefficient μ is a measurable form of type $(-1, 1)$ on X; such a form defines in a known way a conformal structure on the surface X^μ. There is a natural Teichmüller metric

$$\tau_T([X^\mu], [X^\nu]) = \tfrac{1}{2} \inf \log k(W^\mu \circ (W^\nu)^{-1}), \tag{4}$$

on $T(X)$, where μ and ν run through the corresponding classes of equivalent structures, and $k(W^\mu) = (1+\|\mu\|_\infty)/(1-\|\mu\|_\infty)$ ($\|\cdot\|_\infty$ is the sup-norm). In the sequel we deal only with hyperbolic surfaces.

The structure of a complex Banach manifold is introduced on $T(X)$. For this structure X is uniformized by a torsion-free Fuchsian group Γ with invariant unit circle; then the Beltrami differentials μ lift from X to Δ and we continue them by zero to Δ^*. By the same token, quasiconformal homeomorphisms of X are lifted to quasiconformal automorphisms W^μ of the sphere $\widehat{\mathbb{C}}$ with

$$\mu \in L_\infty(\Delta, \Gamma)_1 := \{\nu \in L_\infty(\mathbb{C}) : \nu|_{\Delta^*} = 0,\ \|\nu\|_\infty < 1,$$
$$(\nu \circ \gamma)\overline{\gamma'}/\gamma' = \nu \text{ for all } \gamma \in \Gamma\};$$

we normalize them by the condition $W^\mu(z) = z + O(z^{-1})$ as $z \to \infty$. To the space $T(X)$ corresponds canonically the Teichmüller space $T(\Gamma)$ of the group Γ, obtained by identifying the elements $\mu, \nu \in L_\infty(\Delta, \Gamma)_1$, for which $W^\mu|_{\overline{\Delta^*}} = W^\nu|_{\overline{\Delta^*}}$. Its complex structure is inherited from the ball $L_\infty(\Delta, \Gamma)_1$; in this connection the projection $\phi: L_\infty(\Delta, \Gamma)_1 \to T(\Gamma)$ is holomorphic.

Passing now to the Schwarzian derivative

$$S_{W^\mu}(z) = \left(\frac{W''}{W'}\right)' - \frac{1}{2}\left(\frac{W''}{W'}\right)^2, \qquad z \in \Delta^*,$$

we obtain a biholomorphic imbedding of $T(\Gamma)$ (and $T(X)$) into the Banach space $B_2(\Delta^*, \Gamma)$ of holomorphic Γ-automorphic forms φ in Δ of weight -4 (i.e., $(\varphi \circ \gamma)\gamma'^2 = \varphi$, $\gamma \in \Gamma$; $\varphi(z) = o(|z|^{-4})$ as $z \to \infty$) with norm

$$\|\varphi\| = \sup_{\Delta^*}(|z|^2 - 1)^2|\varphi(z)|.$$

The image of $T(\Gamma)$ under this imbedding is a bounded region lying in the ball $\{\|\varphi\| < 6\}$; often one identifies it with $T(\Gamma)$.

In particular, for $\Gamma = 1$ we have the universal Teichmüller space $T(1) = T$; all $T(\Gamma)$ imbed in it in a natural manner.

The metric (4) is Finsler; the Finsler form F generating it is defined on the tangent bundle $T(T(\Gamma))$ by the formula

$$F(\phi(\mu), \phi'(\mu)\nu) = \inf\{\|\nu^*(1 - |\mu|^2)^{-1}\|_\infty : \phi'(\mu)\nu^* = \phi'(\mu)\nu; \\ \mu \in L_\infty(\Delta, \Gamma)_1; \ \nu, \nu^* \in L_\infty(\Delta, \Gamma)\}. \tag{5}$$

The Kobayashi metric d_T on $T(\Gamma)$ is, as usual, the largest pseudometric $d(\cdot, \cdot)$ on $T(\Gamma)$ for which $d(h(z'), h(z'')) \leq \rho(z', z'')$ for all $h \in \mathrm{Hol}(\Delta, T(\Gamma))$, where ρ is the hyperbolic metric on the disk Δ of curvature -4 (i.e., with differential element $d\rho = |dz|/(1 - |z|^2)$).

A detailed account of these questions can be found, for example, in the books [4] and [11].

§5

We now turn to the proof of our theorems. We first make essential use of the important theorem of Royden-Gardiner on the coincidence of the Kobayashi and Teichmüller metrics [14], [4].

PROPOSITION 1 (Royden-Gardiner). *For each Teichmüller space $T(\Gamma)$ its Kobayashi metric d_T coincides with the Teichmüller metric τ_T, and*

$$d_T(x, y) = \tau_T(x, y) = \inf\{\rho(h^{-1}(x), h^{-1}(y)) : h \in \mathrm{Hol}(\Delta, T(\Gamma)\}.$$

The next assertion concerns properties of holomorphic mappings of the disk with isolated zeros.

PROPOSITION 2. *Let $h: \Delta \to T(\Gamma)$ be a holomorphic mapping such that*

$$h'(0) = h''(0) = \cdots = h^{(m)}(0) = 0, \quad h^{(m+1)}(0) \neq 0, \quad m \geq 1.$$

Then the inequality

$$d_T(h(0), h(t)) \leq \rho(0, |t|^{m+1}) \tag{6}$$

holds in the disk Δ.

The proof of this assertion, which is a generalization of the usual Schwarz lemma for bounded functions with a zero at the origin, is quite long and is based on a modification of the construction lying at the foundation of the proof of Royden and Gardiner on the coincidence of the metrics d_T and

τ_T, and utilizes properties of the Finsler structure of $T(\Gamma)$; it is carried out in several stages.

First, Proposition 2 is established for finite-dimensional Teichmüller spaces $T(\Gamma)$ corresponding to finitely generated groups Γ of the first kind.

We consider the Finsler form (5). We need to estimate the values of this form on the holomorphic disk $h(\Delta) \subset T(\Gamma)$ or, in other words, to estimate the conformal metric $\lambda(t)\,|dt|$ induced by this form on the disk Δ, with $\lambda(t) = F(h(t), h'(t))$. For this we compare it with the values

$$F\left(t^{m+1}\frac{\overline{\varphi}}{|\varphi|}, (m+1)t^m\frac{\overline{\varphi}}{|\varphi|}\right),$$

where φ is the corresponding integrable holomorphic quadratic differential on Δ/Γ, defined by h. From general properties of extremal quasiconformal mappings, we obtain

$$F\left(t^{m+1}\frac{\overline{\varphi}}{|\varphi|}, (m+1)t^m\frac{\overline{\varphi}}{|\varphi|}\right) = \frac{(m+1)|t|^m}{1-|t|^{2(m+1)}} =: \lambda_m(t). \tag{7}$$

We note that the metric $\lambda_m\,|dt|$ is C^2-smooth on $\Delta\backslash\{0\}$ and of constant curvature -4. It plays a critical role in the class of metrics with a zero of order m at the origin, of curvature no greater than -4.

Changing as necessary the scheme of reasoning from [14] and [4] with regard to the properties of the form F, we establish that

(i) $F(h(t), h'(t)) \geq \lambda_m(t) + o(t^{m+2})$

(from this, λ_m serves as a support metric of λ).

(ii) $\lambda(t) = (m+1)|t|^m + o(t^{m+1})$ as $t \to 0$.

Thenthe following simple generalization of the classical lemma of Schwarz-Ahlfors is used:

LEMMA. *Let a continuous conformal metric $\lambda(t)\,|dt|$ be given in the disk Δ, with a zero of order $m \geq 1$ at the origin, where*

$$\lambda(t) = c|t|^m + o(t^m) \quad \text{as } t \to 0 \text{ with } c \leq m+1,$$

admitting at each point $t_0 \in \Delta\backslash\{0\}$, where $\lambda \neq 0$, a C^2-smooth support metric of curvature no larger than -4. Then $\lambda(t) \leq \lambda_m(t)$ for all $t \in \Delta$.

Now the required inequality (6) is obtained from (7) and (i) directly by integration. For infinite-dimensional spaces $T(\Gamma)$ the Gardiner approximation by finite-dimensional spaces is constructed.

§6

We now consider the Schwarzian derivative S_f of a function $f \in \Sigma$ in Δ^*. The correspondence

$$t \mapsto S_{f_t} = S_f(z/t)t^{-2}$$

defines a holomorphic function $h(t)$ from Δ to the universal Teichmüller space T, modeled as a bounded region in $B_2(\Delta^*, \mathbb{1})$; in addition $h'(0) = 0$, $h''(0) \neq 0$.

Theorem 1 is obtained by an application of Propositions 1 and 2. Namely, on the basis of the holomorphic contractibility of the Kobayashi metric it follows from these propositions that

$$\tau_T(0, h(t)) = d_T(0, h(t)) \leq \rho(0, |t|^2),$$

and this means that $f_t \in \Sigma_{|t|^2}$.

To obtain the second assertion of Theorem 1 it is still necessary to use the relation between d_T and the Kobayashi metric $d_{T,k}$ of the hyperbolic ball $B_T(0, k) = \{x \in T(\Gamma)\colon d_T(X, 0) < k\}$.

The proof of Theorem 2 is carried out according to several different schemes, with preservation of the same basic idea. The presence of the supplementary normalization at zero requires, roughly speaking, that one use the Teichmüller space of the punctured disk rather than the universal Teichmüller space, but then the structure of the set filled in $B_2(\Delta^*, \mathbb{1})$ by the corresponding Schwarzian derivatives S_f is unknown. Therefore one has to follow a different path.

An approximation of mappings that is close to the Gardiner approximation of infinite-dimensional Teichmüller spaces by finite-dimensional ones is constructed, and a topological theorem of Epstein on homotopy with fixed base point is used. Correspondingly $t \mapsto f_t \in \Sigma_k(0)$ allows one to define a holomorphic mapping from Δ to the n-dimensional Teichmüller space $T(0, n+3)$ of spheres with $n+3$ punctures $(n \geq 1)$, more precisely, into the balls $B(0, k) \subset T(0, n+3)$. To these mappings is applied the corresponding inequality for the metric $d_{T,k}$ and then passage to the limit as $n \to \infty$ is studied.

It is sufficient to establish the assertion of Theorem 3 for those conformal mappings f that admit a quasiconformal extension $\hat{f}$ into the exterior of the annulus. Then $\hat{f} = W_2 \circ W_1$, where W_1 and W_2 are conformal respectively in $\{|z| > r\}$ and $W_1(|z| < R)$. In W_1 Theorem 1 is directly applicable, but in W_2 the interpolating family is constructed in a more complicated way. Using Strebel's condition [16] for the so-called framed mappings, we get that W_1 admits a k_1-quasiconformal extension W^{μ_1} in Δ with Beltrami coefficients $\mu_1 = k_1\overline{\varphi}_1/|\varphi_1|$, and W_2 admits a k_2-quasiconformal extension W^{μ_2} in $W_1(\Delta^*)$ with $\mu_2 = k\overline{\varphi}_2/|\varphi_2|$, where $k_1 \leq r^2$, $k_2 \leq 1/R^2$, and φ_1, φ_2 are holomorphic and integrable respectively in Δ and $W_1(\Delta^*)$. The unknown reflection (image) with respect to these extensions is obtained by composition with conformal mappings, from which the estimate (2) is established. A detailed proof is presented in [9] and [10].

For functions and curves with supplementary properties, for example of symmetry type, it is possible to extract stronger results. In particular, in case of p-tuple symmetry of rotation around the origin of coordinates, i.e., for

$$f(ze^{2l\pi i/p}) = f(z)e^{2e\pi i/p}, \qquad l = 0, 1, \dots, p-1,$$

the exponent 2 in Theorems 1 and 3 is replaced by p.

As an example we consider the lemniscate

$$L_R = \{W : |P(W)| = R^n > 1, \ P(W) = (W - W_1)\cdots(W - W_n)\} \qquad (8)$$

assuming the polynomial P is such that the exterior of the curve L_1, containing infinity, is simply-connected; in this connection the roots W_j are not necessarily all different. Then it is possible to select in $\operatorname{ext} L_1$ a single-valued branch of the function $z = \sqrt{P(W)}$ with expansion

$$z = W + a_0 + a_1 W^{-1} + \cdots$$

that maps this region conformally onto Δ^*. In addition, L_R is the inverse image of the circle $\{|z| = R\}$, and by Theorem 1, $q_{L_R} \le 1/R^2$. If $P(W) = W^n - 1$, then $q_{L_R} \le 1/R^{2n}$. Clearly, these estimates are crude, but in the general case they are the best at the present time.

§7

Thus, in the case of analytic curves that are level curves of conformal mappings, the problem is completely solved. What can be said for other quasiconformal curves?

Let the curve L be star-shaped with respect to the origin; then again it is possible to examine the curves $L_r = rL$ in $\operatorname{ext} L$ (and $\operatorname{int} L$) and the corresponding interpolating family $rf(z/r)$, $0 < r < 1$, for holomorphic functions $f(z) = z + a_0 + o(z^{-1})$, but now it is impossible to pass to the disk $\{|r| < 1\}$. We normalize the family L_r so that $L = L_1$ has exterior conformal radius at infinity equal to 1, i.e., $\operatorname{ext} L$ is mapped conformally onto Δ^* by a function $g(z) = z + o(z^{-1})$.

In order to estimate $k(f_r)$ it is necessary to be able to separate the curves L_r by level curves Λ_ρ of the function g, i.e., to know an estimate of r by ρ (then $qf(L_r)$ is estimated by $k(f_r)$ and q_{L_r}). But there is no such estimate in the general case.

There is another way—one can use the lemniscate (8) to separate L_r from L_1, taking into account that, according to a theorem of Hilbert, an entire closed Jordan curve can be approximated from the outside by lemniscates (see [5], [17]). But here again it is necessary to have quantitative estimates of approximation.

Some estimates for the reflection coefficient for concrete curves can be found in [13].

REFERENCES

1. J. Becker, *Löwnersche Differentialgleichung und quasikonform fortsetzbare schlichte Funktionen*, J. Reine Angew. Math. **255** (1972), 23–43.

2. D. M. Campbell, J. G. Clunie, and W. K. Hayman, *Research problem in complex analysis*, in: Aspects of Contemporary Complex Analysis, Academic Press, London, 1980, pp. 527–571.

3. D. Gaier, *Konstruktive Methoden der konformen Abbildung*, Springer Tracts in Natural Philosophy, vol. 3, Springer-Verlag, Berlin, 1964.

4. F. P. Gardiner, *Teichmüller theory and quadratic differentials*, Wiley-Interscience, New York, 1987.

5. D. Hilbert, *Über die Entwicklung einer beliebigen analytischen Funktionen einer Variablen in eine unendliche nach ganzen rationalen Funktionen fortschreitende Reihe*, Göttinger Nachr., 1897, pp. 63–70.

6. O. Hübner, *Die Faktorisierung konformer Abbildungen und Anwendungen*, Math. Z. **92** (1966), no. 2, 95–109.

7. S. L. Krushkal′, *Invariant metrics on Teichmüller spaces and quasiconformal extendability of analytic functions*, Ann. Acad. Sci. Fenn. Ser. A I Math. **10** (1985), 299–303.

8. ——, *A new approach to variational problems of the theory of quasiconformal mappings*, Dokl. Akad. Nauk SSSR **292** (1987), 1297–1300; Engish transl. in Soviet Math. Dokl. **35** (1987).

9. ——, *The interpolating family of a univalent analytic function*, Sibirsk. Mat. Zh. **28** (1987), no. 5, 88–94; English transl. in Siberian Math. J. **28** (1987), no. 5.

10. ——, *Continuation of conformal mappings, and hyperbolic metrics*, Sibirsk Mat. Zh. **30** (1989), no. 5, 84–100; English transl. in Siberian Math. J. **30** (1989), no. 5.

11. S. L. Kruschkal′ [Krushkal] and R. Kühnau, *Quasiconformal mappings—new methods and applications*, Teubner Texts in Mathematics, vol. 54, Teubner, Leipzig, 1983.

12. R. Kühnau, *When are the Grunsky coefficient conditions sufficient for Q-quasiconformal continuability?*, Comment. Math. Helv. **61** (1985), no. 2, 290–307.

13. ——, *Extremal conformal reflection in a Jordan curve*, Jahresber, Deutsch. Math.-Verein **90** (1988), no. 2, 90–109.

14. H. L. Royden, *Automorphisms and isometrics of Teichmüller spaces*, Ann. of Math. Stud. **66** (1971), 369–383.

15. M. Schiffer, *Fredholm eigenvalues and Grunsky matrices*, Ann. Polon. Math. **39** (1981), 149–164.

16. K. Strebel, *On the existence of extremal Teichmüller mappings*, J. Analyse Math. **30** (1976), 464–480.

17. J. L. Walsh, *Interpolation and approximation by rational functions in the complex domain*, 2nd ed., Amer. Math. Soc. Colloq. Publ., vol. 20, Amer. Math. Soc., Providence, RI, 1956.

Translated by A. L. MATHESON

Amer. Math. Soc. Transl.
(2) Vol. **151**, 1992

Nonstandard Methods in Geometric Functional Analysis

UDC 517.11–517.98

A. G. KUSRAEV AND S. S. KUTATELADZE

Nonstandard methods in the modern sense consist of the explicit or implicit appeal to two different models of set theory—"standard" and "nonstandard"—to investigate concrete mathematical objects and problems. The main development of such methods dates to the last thirty years, and they have now crystallized in several directions (see [29], [42] and the bibliography cited there). The main directions are now known as infinitesimal and Boolean analysis. In this paper we shall outline new applications of nonstandard methods to problems arising in the area of our personal interests, grouped together under the general heading of geometric functional analysis [48]; we shall also point out some promising directions of further research.

§1. Infinitesimal analysis

1.1. Infinitesimal analysis, following its creator A. Robinson, is frequently referred to by the expressive but rather unfortunate phrase "nonstandard analysis"; nowadays one most frequently speaks of classical or Robinsonian nonstandard analysis. Infinitesimal analysis is characterized by the use of certain conceptions, long familiar in the practice of natural sciences but frowned upon in twentieth-century mathematics, involving the notions of actual infinitely large and infinitely small quantities.

1.2. Modern expositions of nonstandard analysis rely on formulas of E. Nelson's internal set theory IST [58] and its later developments, the external set theories of K. Hrbacek (EXT) [49] and T. Kawai (NST) [53]. From the standpoint of the "working mathematician-Philistine," the essence of these theories is as follows.

1991 *Mathematics Subject Classification*. Primary 46S20, 03H05.

Ordinary mathematical objects and properties are called internal (and considered, if a rigorous formalization is desired, within the framework of Zermelo-Fraenkel set theory ZFC). One introduces a new predicate $\mathrm{St}(x)$, expressing the property of an object x to be standard (qualitatively speaking—obtained through existence and uniqueness theorems, i.e., the set of natural numbers is standard, but the infinitely large natural numbers are nonstandard). Mathematical formulas and concepts in whose construction the new predicate is used will be called external. "Cantorian" sets endowed with external properties are referred to as external. In Nelson's theory, such sets are considered only as terms of a metalanguage, which is used only for convenience. In EXT and NST one can treat them as objects of Zermelo theory, which requires elaboration of a formalism and introduction of a new primary predicate $\mathrm{Int}(x)$, stating that the object x is internal. The available formalisms ensure that the extension of ZFC is conservative, i.e., when proving mathematical statements whose formulations do not involve external concepts, we may legitimately invoke the theories IST, EXT, and NST, as no less reliable than ZFC.

1.3. A point of crucial importance is that the new theories contain additional rules, easily motivated at the intuitive level, which are known as the principles of nonstandard analysis. We present their rigorous formulations in IST.

(1) Transfer principle:

$$(\forall^{\mathrm{st}} x_1)\cdots(\forall^{\mathrm{st}} x_n)((\forall^{\mathrm{st}} x)\varphi(x, x_1, \dots, x_n) \to (\forall x)\varphi(x, x_1, \dots, x_n)),$$

where φ is an internal formula and $\varphi = \varphi(x, x_1, \dots, x_n)$ (i.e., φ does not contain any free variables other than those listed).

(2) Idealization principle:

$$(\forall x_1)\cdots(\forall x_n)(\forall^{\mathrm{st\,fin}} z)(\exists x)(\forall y \in z)\varphi(x, y, x_1, \dots, x_n)$$
$$\leftrightarrow (\exists x)(\forall^{\mathrm{st}} y)\varphi(x, y, x_1, \dots, x_n),$$

where φ is an internal formula and $\varphi = \varphi(x, y, x_1, \dots, x_n)$.

(3) Standardization principle:

$$(\forall x_1)\cdots(\forall x_n)(\forall^{\mathrm{st}} x)(\exists^{\mathrm{st}} y)(\forall^{\mathrm{st}} z) z \in y \leftrightarrow z \in x \wedge \varphi(z, x_1, \dots, x_n),$$

where $\varphi = \varphi(z, x_1, \dots, x_n)$ is an arbitrary formula. The index st indicates that the quantifier in question is relativized to standard sets; the index st fin has the analogous meaning with regard to standard finite sets.

§2. Boolean-valued analysis

2.1. Boolean-valued analysis is characterized by the extensive use of the terms lowering and lifting, cyclic hulls and mixing. The development of this trend, which emerged under the impetus of P. J. Cohen's remarkable work on the continuum hypothesis, leads to essentially new ideas and results, first and foremost, in the theory of Kantorovich spaces and von Neumann algebras. The modeling device offered by Boolean-valued analysis makes it possible, in particular, to consider the elements of functional classes as numbers, which substantially facilitates the analysis and creates a unique possibility of automatically extending the scope of classical theorems.

2.2. The construction of a Boolean-valued model begins with a complete Boolean algebra B. For every ordinal $\alpha \in O_n$ one defines

$$V_\alpha^{(B)} := \{x : (\exists \beta \in \alpha) x : \mathrm{dom}(x) \to B \wedge \mathrm{dom}(x) \in V_\beta^{(B)}\}.$$

After this recursive definition, one introduces the Boolean-valued universe $V^{(B)}$ or class of B-sets:

$$V^{(B)} := \bigcup_{\alpha \in O_n} V_\alpha^{(B)}.$$

2.3. Taking an arbitrary formula of ZFC and interpreting the connectives and quantifiers in the natural way in the Boolean algebra B, one defines its truth value $[\![\varphi]\!]$, which depends on the way in which φ is built up from atomic formulas $x = y$ and $x \in y$. The truth values of the latter are defined for $x, y \in V^{(B)}$ by a recursion schema:

$$[\![x \in y]\!] := \bigvee_{z \in \mathrm{dom}(y)} y(z) \wedge [\![z = x]\!],$$

$$[\![x = y]\!] := \bigvee_{z \in \mathrm{dom}(x)} x(z) \Rightarrow [\![z \in y]\!] \wedge \bigwedge_{z \in \mathrm{dom}(y)} y(z) \Rightarrow [\![z \in x]\!]$$

(the sign $\Rightarrow$ symbolizes implication in B).

The universe $V^{(B)}$ with the above valuation rule is a ("nonstandard") model of set theory in the following sense.

2.4. Transfer principle. For any theorem φ of ZFC, the formula $[\![\varphi]\!] = 1$ is valid, i.e., φ is true inside $V^{(B)}$.

2.5. In the class $V^{(B)}$ there is a natural equivalence $x \sim y := [\![x = y]\!] = 1$, which preserves truth values. In this connection, one can use a special device to go over to a separated universe $\overline{V}^{(B)}$, in which $x = y \leftrightarrow [\![x = y]\!] = 1$. In fact, the identification $V^{(B)} := \overline{V}^{(B)}$ is usually assumed without special mention. The basic properties of $V^{(B)}$ are expressed by the following assertions.

2.6. Mixing principle. Let $(b_\xi)_{\xi \in \Xi}$ be a partition of unity in B, i.e., $\xi \neq \eta \to b_\xi \wedge b_n = 0$, $V_{\xi \in \Xi} b_\xi = 1$. For any family $(x_\xi)_{\xi \in \Xi}$ of the universe $V^{(B)}$ there exists a (unique) mixture of $(x_\xi)_{\xi \in \Xi}$ with probabilities $(b_\xi)_{\xi \in \Xi}$, i.e., an element x of the separated universe, denoted by $\sum_{\xi \in \Xi} b_\xi x_\xi$ or $\mathrm{mix}_{\xi \in Z} b_\xi x_\xi$, such that $[\![x = x_\xi]\!] \geq b_\xi$ for $\xi \in \Xi$.

2.7. Maximum principle. For every formula φ of ZFC there is an element $x_0 \in V^{(B)}$ for which

$$[\![(\exists x)\varphi(x)]\!] = [\![\varphi(x_0)]\!].$$

In particular, $V^{(B)}$ contains an object $\mathscr{R}$ which plays the role of the field R inside $V^{(B)}$.

2.8. Besides the above principles, there is an important procedure for passing to $V^{(B)}$ from the ordinary von Neumann universe V, where the latter is defined by the recursion schema

$$V_\alpha := \{x: (\exists \beta \in \alpha) x \in P(V_\beta)\}, \qquad V := \bigcup_{\alpha \in O_n} V_\alpha.$$

This procedure is defined by the rule

$$\varnothing^\wedge := \varnothing, \quad \operatorname{dom}(x^\wedge) := \{y^\wedge : y \in x\}, \quad \operatorname{im}(x^\wedge) := \{1\}.$$

The element $x^\wedge \in V^{(B)}$ is known as the standard name of x. We thus have a canonical embedding of V into $V^{(B)}$. Apart from this we have a technique of lowerings and liftings of sets and correspondences.

2.9. Given an element $x \in V^{(B)}$, its lowering $x \downarrow$ is defined by the rule $x \downarrow := \{t \in V^{(B)} : [\![t \in x]\!] = 1\}$. The set $x \downarrow$ is cyclic, i.e., closed with respect to mixing of its elements.

2.10. Let F be a correspondence from X to Y inside $V^{(B)}$. There exists a correspondence $F \downarrow$ —and it is unique— from $X \downarrow$ to $Y \downarrow$ such that for any subset A of X inside $V^{(B)}$, we have $F(A) \downarrow = F \downarrow (A \downarrow)$.

In particular, a map $f: X^\wedge \to Y$ inside $V^{(B)}$ defines a function-lowering $f \downarrow : X \to Y \downarrow$ such that $f \downarrow (x) = f(x^\wedge)$ $(x \in X)$.

2.11. Let $x \in P(V^{(B)})$. Define $\varnothing \uparrow := \varnothing$ and $\operatorname{dom}(x \uparrow) = x$, $\operatorname{im}(x \uparrow) = \{1\}$. The element $x \uparrow$ is called the lifting of x. It is easy to see that $x \uparrow\downarrow$ is the least cyclic set containing x, i.e., its cyclic hull: $x \uparrow\downarrow = \operatorname{mix}(x)$.

2.12. Let $X, Y \in P(V^{(B)})$ and let F be a correspondence form X to Y. There exists a correspondence $F \uparrow$ —and it is unique— from $X \uparrow$ to $Y \uparrow$ inside $V^{(B)}$ such that $\operatorname{dom}(F \uparrow) = \operatorname{dom}(F) \uparrow$ and for every subset A of $\operatorname{dom}(F)$ we have $F \uparrow (A \uparrow) = F(A) \uparrow$ if and only if F is extensional, i.e.,

$$y_1 \in F(x_1) \to [\![x_1 = x_2]\!] \leq \bigvee_{y_2 \in F(x_2)} [\![y_1 = y_2]\!].$$

In particular, a map $f: X \to Y \downarrow$ generates a function $f \uparrow : X^\wedge \to Y$ such that $f \uparrow (x^\wedge) = f(x)$ for $x \in X$. If necessary in specific cases, the lowering and lifting procedures can be iterated.

§3. Vector lattices

3.1. There are several excellent monographs on the theory of vector lattices [4], [18], [19], [55], [70]. Vector lattices are also commonly known as Riesz spaces, and order-complete vector lattices as Kantorovich spaces or K-spaces. A K-space is said to be extended if any set of pairwise disjoint positive elements in it has a supremum. The most important examples of extended K-spaces are the following:

(1) the space $M(\Omega, \Sigma, \mu)$ of equivalence classes of measurable functions, where (Ω, Σ, μ) is a measure space with μ a σ-finite measure (or, more generally, a space with the direct sum property, see [18]);

(2) the space $C_\infty(Q)$ of continuous functions defined on an extremally disconnected compact space Q with values in the extended real line, taking the values $\pm\infty$ only on a nowhere dense set [4], [19], [55];

(3) the space $\overline{A}$ of selfadjoint (not necessarily bounded) operators associated with a von Neumann algebra (see [66]).

To save space, we shall restrict attention to the real case, since the analysis of complex K-spaces is entirely analogous. The symbol $\mathfrak{P}(E)$ will denote the Boolean algebra of order projections in a K-space E. If E contains an order unit, $\mathfrak{E}(E)$ is the Boolean algebra of unit elements (fragments of the identity) in E. The algebras $\mathfrak{P}(E)$ and $\mathfrak{E}(E)$ are isomorphic and known as the base of E. Throughout the sequel, B will be a fixed complete Boolean algebra. The basis for Boolean-valued analysis of vector lattices is the following result.

3.2. Theorem (Gordon [6]). *Let $\mathscr{R}$ be the field of real numbers in the model $V^{(B)}$. The algebraic system $\mathscr{R}\downarrow$ (i.e., $\mathscr{R}$ with lowered operations and order) is an extended K-space. Moreover, there exists an isomorphism χ of the Boolean algebra B onto the base $\mathfrak{P}(E)$ such that*

$$b \le [\![x = y]\!] \leftrightarrow \chi(b)x = \chi(b)y,$$
$$b \le [\![x \le y]\!] \leftrightarrow \chi(b)x \le \chi(b)y$$

for all $x, y \in \mathscr{R}\downarrow$ and $b \in B$.

Throughout the sequel, R will denote the field of real numbers inside $V^{(B)}$. If the base of a K-space E is isomorphic to B, then E itself is isomorphic to the foundation $E_0 \subset \mathscr{R}\downarrow$, and in this situation E is extended if and only if $E_0 = \mathscr{R}\downarrow$. Under these circumstances one says that $\mathscr{R}\downarrow$ is a maximal extension and $\mathscr{R}$ a Boolean-valued realization of the K-space E. It is noteworthy that Boolean-valued realizations of certain structures lead to subsystems of the field $\mathscr{R}$.

3.3. Theorem [25].

(1) *A subgroup of the additive group of $\mathscr{R}$ is a Boolean-valued realization of an archimedean lattice-ordered group.*

(2) *A vector sublattice of $\mathscr{R}$, considered as a vector lattice over the field $R^\wedge$ is a Boolean-valued realization of an archimedean vector lattice.*

(3) *An archimedean f-ring contains two mutually complementary components, one of which is a group with zero multiplication realized as in* (1), *and the other has a subring of the ring $\mathscr{R}$ as a Boolean-valued realization.*

(4) *An archimedean f-algebra contains two mutually complementary components, one of which is a vector lattice with zero multiplication realized as in*

(2), *and the other is realized as a subring and sublattice of* $\mathscr{R}$, *considered as an* f-*algebra over* $R^{\wedge}$.

3.4. Gordon's theorem implies the main structural properties of K-spaces. We shall dwell on the realization of K-spaces and functional calculus. Let Q be a Stonean compact subspace of the Boolean algebra B and define $C_{\infty}(Q)$ as in 3.1 (2). We call a map $e: R \to B$ a resolution of unity in B if (1) $e(s) \le e(t)$ $(s \le t)$; (2) $\bigvee_{t\in R} e(t) = 1$, $\bigwedge_{t\in R} e(t) = 0$; (3) $\bigvee_{s<t} e(s) = e(t)$ $(t \in R)$. Let $B(R)$ be the set of all resolutions of unity in B. The sets $C_{\infty}(Q)$ and $B(R)$ can be endowed canonically with the structure of an extended K-space (see [4] and [19]).

3.5. Theorem ([29], [50]). *The extended* K-*space* $\mathscr{R}\downarrow$ *is (algebraically and order) isomorphic to each of the* K-*spaces* $B(R)$ *and* $C_{\infty}(Q)$. *Under this isomorphism an element* $x \in \mathscr{R}\downarrow$ *is mapped onto a resolution of unity* $t \to e_t^x$ $(t \in R)$ *and onto a function* $\overline{x}: Q \to \overline{R}$ *by the formulas*

$$e_t^x := [\![x < t^{\wedge}]\!] \qquad (t \in R),$$
$$\overline{x}(q) := \inf\{t \in R: [\![x < t^{\wedge}]\!] \in q\} \qquad (q \in Q).$$

3.6. Let $\mathscr{B}_R$ and $\mathscr{B}(R)$ be the σ-algebra of Borel sets and the vector lattice of Borel functions, respectively, on the real line. We identify B with the algebra of fragments of the identity in $\mathscr{R}\downarrow$ (see 3.2). For every $x \in \mathscr{R}\downarrow$ there exists a unique spectra measure (= sequentially o-continuous Boolean homomorphism $\mu: \mathscr{B}_R \to B$) such that $\mu(-\infty, t) = e_t^x$ $(t \in R)$. The measure μ defines an integral

$$I_x(f) := \int_R f(t)\, d\mu(t) \qquad (f \in \mathscr{B}(R)).$$

In this situation $I_x(f)$ is the unique element of $\mathfrak{R}\downarrow$ for which

$$[\![I_x(f) < t^{\wedge}]\!] = \mu(\{f < t\}).$$

3.7. Theorem ([29], [50]). *The map* $I_x: \mathscr{B}(R) \to \mathscr{R}\downarrow$ *is the unique sequentially* o-*continuous lattice and algebraic homomorphism satisfying the condition*

$$I_x(\mathrm{id}_R) = x.$$

3.8. For other aspects of Boolean-valued analysis of vector lattices, see [7], [8], [24], [29], [50], [51], [65].

§4. Positive operators

4.1. General information about positive and order-bounded operators may be found in [24], [29]. Take arbitrary K-spaces Z and E. A positive operator $\Phi: Z \to E$ will be called a Maharam operator if it is order continuous and $\Phi([0, z]) = [0, \Phi(z)]$ for every $z \in Z^{+}$, where $[a, b] := \{c: a \le c \le b\}$ is an order interval. Let mZ be a maximal extension of Z and $D(\Phi)^{+}$ the set of all $0 \leq z \in mZ$ such that $\{\Phi z' : z' \in Z,\ 0 \leq z' \leq z\}$ is bounded. Then $D(\Phi) := D(\Phi)^+ - D(\Phi)^+$ is a foundation in mZ and Φ extends to a Maharam operator on the whole of $D(\Phi)$. We say that Φ is essentially positive if $\Phi \geq 0$ and $\Phi(|z|) = 0$ implies $z = 0$.

4.2. Theorem [22]. *Let Φ be an essentially positive Maharam operator. There exist a K-space $\mathcal{Z}$ and an essentially positive o-continuous functional $\varphi : \mathcal{Z} \to \mathcal{R}$ in the model $V^{(B)}$, and there exists an isomorphism ι from $D(\Phi)$ onto the K-space $\mathcal{Z} \downarrow$ such that $\Phi = \varphi \downarrow \circ \iota$.*

4.3. The above result reduces the investigation of Maharam operators to analysis of the class of o-continuous positive functionals. What is the situation with regard to arbitrary positive operators? Various approaches based on Boolean-valued analysis may be adopted here. Let us consider an order-bounded operator from a vector lattice Z into $E : \mathcal{R} \downarrow$. There exists an order-bounded $R^\wedge$-linear functional $\varphi : Z^\wedge \to R$ inside $V^{(B)}$ for which $\Phi = \varphi \downarrow \circ j$, where $j : z \to z^\wedge$ $(z \in Z)$. The map $\Phi \to \varphi$ is an isomorphism of the space of all order-bounded operators $L_r(Z, E)$ onto $\widetilde{\mathcal{Z}} \downarrow$, where $\widetilde{\mathcal{Z}}$ is the space of order-bounded functionals on $\mathcal{Z}$. In particular, $\Phi \geq 0$ if and only if $[\![\varphi \geq 0]\!] = 1$. The disadvantage of this device is that the map $\Phi \to \varphi$ does not preserve order-continuity.

On the other hand, for a positive operator $\Phi : Z \to E$ one can construct an essentially positive Maharam operator $\overline{\Phi}$ and a lattice homomorphism $h : Z \to D(\overline{\Phi})$ such that $\Phi = \overline{\Phi} \circ h$, where the pair $(h, \overline{\Phi})$ is minimal in a certain sense (see [1]). Appealing to Theorem 4.2, we obtain a representation $\Phi = \varphi \downarrow \circ \iota'$, where $\iota' := h \circ \iota$ and φ is an essentially positive o-continuous functional in the model $V^{(B)}$. The disadvantage of this approach is that the space $D(\overline{\Phi})$ may prove to be invisible. However, in a fairly general situation, $D(\overline{\Phi})$ is realized as the space of functions (in two variables) on $P \times Q$, where P and Q are Stonean compact subspaces of Z and E, respectively (see [55]).

4.4. The above arguments are easily applied to the algebra of fragments of an arbitrary positive operator Φ acting from a vector lattice Z to a K-space E with filter of units ξ and base $\mathfrak{P}(E)$ (see [1] and [39]). We dwell on the representation of the projection S of an operator T onto the component $\{\Phi\}^{dd}$ generated by the operator Φ. Let us call a set of operators $\mathcal{P}$ in $L_r(Z, E)$ a generating set if $\Phi x^+ = \sup\{p\Phi x : p \in \mathcal{P}\}$ for all $x \in Z$. To study interesting fragments by lifting into a Boolean-valued universe, one can reduce everything to the case of functionals. For the latter, using infinitesimal representations, one readily proves that

$$Sx \underline{=} \inf{}^* \{{}^\circ pTx : p^d \Phi x \approx 0,\ p \in \mathcal{P}\},$$
$$Sx \underline{=} \inf{}^* \{{}^\circ Ty : \Phi(x - y) \approx 0,\ 0 \leq y \leq x\},$$

where * is the standardization symbol, $^\circ$ the "standard part" operation, $\approx$

denotes infinite smallness and $\rightharpoonup$ denotes the exactness of the formula, i.e., the attainability of equality.

Interpreting the above nonstandard representations and performing the lowering, one arrives at the following formulas [29]:

$$Sx = \sup_{\varepsilon\in\xi} \inf\{\pi T_y + \pi^d T_x : 0 \leq y \leq x\,,\ \pi \in \mathfrak{P}(E)\,,\ \pi\Phi(x - y) \leq \varepsilon\}\,,$$

$$Sx = \sup_{\varepsilon\in\xi} \inf\{(\pi p)^d T_x : p\Phi_x \leq \varepsilon\,,\ p \in \mathscr{P}\,,\ \pi \in \mathfrak{P}(E)\}.$$

§5. Banach-Kantorovich spaces

5.1. A Banach-Kantorovich space consists of a (real or complex) vector space X, a K-space E, and a vector norm $|\cdot|: X \to E$ such that the following conditions hold: (1) the norm is decomposable, i.e., if $|x| = e_1 + e_2$, where $x \in X$ and $e_1, e_2 \in E^+$, then $x = x_1 + x_2$ and $|x_k| = e_k$ $(k := 1, 2)$ for suitable $x_1, x_2 \in X$; (2) X is o-complete, i.e., for any net $(x_\alpha) \subset X$, if $o\text{-}\lim |x_\alpha - x_\beta| = 0$, then $o\text{-}\lim |x_\alpha - x| = 0$ for some $x \in X$. We shall assume that $\{|x| : x \in X\}^{dd} = E \subset \mathscr{R}\downarrow$. If E is extended, i.e., $E = \mathscr{R}\downarrow$, then X is also said to be extended. An example of an extended Banach-Kantorovich space is the space $M(\Omega, \Sigma, \mu, Y)$ of (equivalence classes of) strongly measurable vector-valued functions with values in a Banach space Y.

5.2. Theorem [23]. *Let x be a Banach space in the model $V^{(B)}$. Then the lowering $x\downarrow$ is an extended Banach-Kantorovich space. Conversely, if X is an extended Banach-Kantorovich space, there exists a unique* (*up to linear isometry*) *Banach space x in $V^{(B)}$ whose lowering is linearly isometric to X.*

5.3. Let us call the bounded part of the space $x\downarrow$ the restricted descent of x. The restricted descents of Banach spaces in $V^{(B)}$ constitute the class of B-cyclic Banach spaces. Let B be the complete Boolean algebra of norm one projections in a Banach space X. We shall say that X is cyclic with respect to B, or B-cyclic, if, for an arbitrary partition of unity $(\pi_\xi)_{\xi\in\Xi} \subset B$ and any bounded family $(x_\xi)_{\xi\in\Xi} \subset X$ there exists a unique element $x \in X$ such that $\pi_\xi x_\xi = \pi_\xi x$ $(\xi \in \Xi)$ and $\|x\| \leq \sup_{\xi\in\Xi} \|x_\xi\|$. Let $A(B)$ denote an arbitrary commutative AW^*-algebra whose complete Boolean algebra of idempotents is isomorphic to B. If X is an AW^*-model over $A(B)$ (see [52]), then X is a B-cyclic Banach space. All the aforesaid leads to the following realization theorem.

5.4. Theorem [59]. *The restricted descent of a complex Hilbert space in the model $V^{(B)}$ is an AW^*-module over the algebra $A(B)$. Conversely, for any AW^*-module X over $A(B)$ there exists a unique* (*up to unitary equivalence*) *Hilbert space inside $V^{(B)}$ whose restricted descent is unitarily equivalent to X.*

5.5. Let X and Y be Banach-Kantorovich spaces with norming lattices E and F, respectively. A linear operator $T: X \to Y$ is said to be majorizable if there exists a positive operator $S: E \to F$ such that $|Tx| \leq S(|x|)$ for all $x \in X$. If $E = F$ and S is an orthomorphism, a majorizable operator is also called bounded, since in that situation T coincides with the lowering from $V^{(B)}$ of a bounded linear operator acting in Banach spaces. By interpreting Riesz-Schauder theory in a Boolean-valued model one arrives at a new concept of cyclic compactness and obtains corresponding results on the solvability of operator equations in Banach-Kantorovich spaces [24]. General majorizable operators have a far more complicated structure and their analysis requires appeal to a considerable variety of methods (see [24], [26], [31]).

5.6. Banach-Kantorovich spaces and majorizable operators were first introduced by L. V. Kantorovich in [16]. It was he who proposed the first applications to the solution of operator equations by the method of successive approximations (see [17], [19]). These objects possess a rich structure and have several important applications in the area of spaces of measurable vector-valued functions and linear operators in such spaces [26]. In particular, the study of Banach-Kantorovich spaces leads to the notion of Banach spaces with mixed norms, which is enormously useful in connection with the isometric classification of Banach function spaces (see [26]).

§6. Banach algebras

6.1. Certain classes of Banach algebras yield some beautiful variations on the theme outlined in the previous section. Call a C^*-algebra A a B-C^*-algebra if A is cyclic with respect to a Boolean algebra of projections B, where any projection in B is multiplicative, involutive and of unit norm. If A is an AW^*-algebra and B a regular subalgebra of the Boolean algebra of central projections $\mathfrak{P}(A)$, then A is a B-C^*-algebra. We shall therefore say that A is a B-AW^*-algebra if B is a regular subalgebra of $\mathfrak{P}_C(A)$. Now let A be a JB-algebra and B and $\mathfrak{P}_C(A)$ the same as before. If A is a cyclic Banach space with respect to B, we shall say that A is a B-JB-algebra. An isomorphism that commutes with the projections in B will be called a B-isomorphism. The following theorem, though in a slightly different form, was proved in [67].

6.2. Theorem [67]. *The restricted descent of a C^*-algebra in the model $V^{(B)}$ is a B-C^*-algebra. Conversely, for every B-C^*-algebra A, there exists inside $V^{(B)}$ a unique (up to $*$-isomorphism) C^*-algebra $\mathscr{A}$ such that the restricted descent of $\mathscr{A}$ is $*$-B-isomorphic to A.*

6.3. Theorem. *The restricted descent of an AW^*-algebra (JB-algebra) from the model $V^{(B)}$ is a B-AW^*-algebra (B-JB-algebra). Conversely, for any B-AW^*-algebra (B-JB-algebra) A there exists a unique (to within an* *isomorphism) AW^*-algebra (JB-algebra) $\mathscr{A}$ whose restricted descent is B-isomorphic to A. In addition, $\mathscr{A}$ will be a factor in $V^{(B)}$ if and only if $B = \mathscr{P}_c(A)$. The formulated statement concerning AW^*-algebras is obtained in* [59] *and* [60].

6.4. The Boolean-valued realization of von Neumann algebras [66] is also worthy of mention. The above realization theorems form the foundation for Boolean-valued analysis of all these classes of Banach algebras (see [59]–[62], [66], [67]). In particular, it was shown in [59] that for all infinite cardinals α and β there exists an AW^*-algebra that is simultaneously α- and β-homogeneous (a conjecture of Kaplansky in [52]). This fact is related to the location of cardinal numbers under embeddings in $V^{(B)}$ (see [44], [68]).

§7. Convex analysis

7.1. The subdifferential is one of the most important concepts in convex analysis (see [24], [28]). In this section, referring to a few examples, we shall show how to use Boolean-valued analysis to study the internal structure of subdifferentials. Take a vector space X, a K-space E, and a sublinear operator $P: X \to E$. The subdifferential ∂P of P at zero is also called the supporting set of P [28]. By Gordon's theorem, we may assume that $E \subset \mathscr{R}\downarrow$, so that we can "convert" P inside a suitable model into an $\mathscr{R}$-valued sublinear operator, i.e., a sublinear functional. To be precise:

7.2. Theorem [54]. *There exist a Banach space x and a continuous sublinear functional $p: x \to \mathscr{R}$ in the model $V^{(B)}$ such that there is an isomorphic embedding of X into the Banach-Kantorovich space $x\downarrow$ with $[\![(\imath X)\uparrow$ is dense in $x]\!] = 1$ and $P = p \circ \imath$. In this situation, for any $U \in \partial P$ there is a unique element $u \in V^{(B)}$ for which $[\![u \in \partial p]\!] = 1$ and $U = u\downarrow \circ \imath$. The map $U \to u$ is an affine isomorphism of the convex sets ∂P and $(\partial p)\downarrow$.*

7.3. Thus, the study of ∂P largely reduces to that of ∂p. For example, let us look at the extremal structure of the subdifferential ∂P. Let $\mathrm{Ch}(P)$ denote the set of extreme points of ∂P. It should be noted that by Theorem 2 the relations $U \in \mathrm{Ch}(P)$ and $[\![u \in \mathrm{Ch}(p)]\!] = 1$ are equivalent, and one can then use the classical Kreĭn-Mil′man Theorem and Mil′man's inversion of it for ∂p. For a rigorous formulation of the result, we need some more definitions. The weak closure $\sigma\text{-}\mathrm{cl}(\Omega)$ (cyclic hull $\mathrm{mix}(\Omega)$) is the set of all operators $T \in L(X, E)$ of the form $Tx = o\text{-}\lim T_\alpha x$ $(x \in X)$, where (T_α) is a net in Ω (resp., $Tx = o\text{-}\sum \pi_\xi T_\xi x$ $(x \in X)$, where $(T_\xi) \subset \Omega$ and (π_ξ) is a partition of unity in $\mathfrak{P}(E)$). The weak cyclic closure of Ω is the set $\sigma\text{-}\mathrm{cl}(\mathrm{mix}(\Omega))$. If $\sigma\text{-}\mathrm{cl}\,\Omega = \Omega$ or $\mathrm{mix}(\Omega) = \Omega$, one says that Ω is weakly o-closed or cyclic, respectively. The definition of weak r-closedness is analogous.

7.4. Theorem [27], [28]. (1) *For any sublinear operator* $P: X \to E$ *the subdifferential coincides with the weakly cyclic closure of the convex hull of its extreme points* $\mathrm{Ch}(P)$.

(2) *If* $P: X \to E$ *is a sublinear operator and* $T \in L(Y, X)$, *then* $\mathrm{Ch}(P \circ T) \subset \mathrm{Ch}(P) \circ T$.

A set $\Omega \subset L(X, E)$ is operator convex (weakly bounded) if $\alpha\Omega + \beta\Omega \subset \Omega$ for any $\alpha, \beta \in \mathscr{R}\downarrow^{+}$, $\alpha + \beta = 1$ (the set $\{T_x : T \in \Omega\}$ is order bounded for all $x \in X$).

7.5. Theorem [24], [27]. *For a weakly bounded set* $\Omega \subset L(X, E)$, *the following assertions are equivalent*:

(1) $\Omega = \partial P$ *for some sublinear* $P: X \to E$;

(2) Ω *is convex, cyclic, and weakly* r*-closed*;

(3) Ω *is convex, cyclic, and weakly* o*-closed*;

(4) Ω *is operator convex and weakly* o*-closed.*

7.6. Let $\Phi: Z \to E$ be a positive operator, P a sublinear operator from a vector space X to a K-space Z. The term disintegration in K-spaces refers to those parts of the calculus of subdifferentials based on the formula $\partial(\Phi \circ P) = \Phi \circ \partial P$. This formula is not always true, but it is known to be valid if Φ is an order-continuous functional $(E = R)$. The general case is analyzed with the help of Theorem 4.2. Let Φ, φ, $\imath$ be the same as in 4.2. There exists an $R^{\wedge}$-sublinear operator $\rho: X^{\wedge} \to \mathscr{Z}$ inside $V^{(B)}$ for which $\rho\downarrow \circ j = \imath \circ P$ (cf. 4.3). From this and 7.2 we conclude that

$$\Phi \circ P = \Phi \circ \imath^{-1} \circ (\imath \circ P) = \varphi\downarrow \circ \rho\downarrow \circ j = (\varphi \circ \rho)\downarrow \circ j,$$
$$\partial(\Phi \circ P) = \{u\downarrow \circ j : [\![u \in \partial(\varphi \circ \rho) = \varphi \circ \partial\rho]\!] = 1\}.$$

These arguments yield the following result.

7.7. Theorem [22]. *Let* Φ *be a positive order-continuous operator. The formula* $\sigma(\Phi \circ P) = \Phi \circ \partial P$ *is valid for any sublinear operator* P *if and only if* Φ *is a Maharam operator.*

7.8. Further developments of disintegration in K-spaces may be found in [24] and [28]. On nonstandard methods in convex analysis see also [33], [34], [36], and [54].

§8. Monadology

8.1. A central concept of infinitesimal analysis is the monad. According to Euclid's definition, "a monad is that through which the many become one." In the formal theory, a monad $\mu(\mathscr{F})$ is defined as an external list of the standard elements of a standard filter $\mathscr{F}$, i.e., $x \in \mu(\mathscr{F}) \leftrightarrow (\forall^{\mathrm{st}} F \in \mathscr{F}) x \in F$. A syntactic characterization of external sets that are monads was proposed not long ago by Benninghofen and Richter [45]. It is useful to emphasize that every monad is a union of ultramonads—monads of ultrafilters. For such a monad U the assertions $(\exists x \in U)\varphi(x)$ and $(\forall x)\varphi(x)$, where $\varphi = \varphi(x)$ is an external formula, are equivalent. Hence it is clear that ultramonads are the genuine "elementary" objects of infinitesimal analysis.

8.2. For applications to the theory of operators, it is of essential importance to construct a synthetic theory in the framework of which both the nonstandard methods offered by Boolean-valued models and external set theories can be used. Only preliminary results have so far been achieved in this direction; they pertain to the study of topological-type notions related with mixing—cyclic filters, topologies and so on, which play major roles in K-spaces. We shall point out one of the possible approaches to cyclic monadology.

8.3. Fix a standard complete Boolean algebra B and an external set A consisting of elements of a separated Boolean-valued universe $V^{(B)}$. An element $x \in V^{(B)}$ is a member of the cyclic hull $\operatorname{mix}(A)$ if and only if, for some internal family $(a_\xi)_{\xi\in\Xi}$ of elements of A and an internal partition of unity $(b_\xi)_{\xi\in\Xi}$ in B, we have
$x = \operatorname{mix}_{\xi\in\Xi} b_\xi x_\xi$. A monad $\mu(\mathscr{F})$ is said to be cyclic if $\mu(\mathscr{F}) = \operatorname{mix}\mu(\mathscr{F})$. A point is said to be essential if it lies in the monad of some pro-ultrafilter—a maximal cyclic filter or, more rigorously, an ultrafilter in $V^{(B)}$.

8.4. Theorem. (1) *A standard filter is cyclic if and only if its monad is cyclic.*

(2) *A filter is extensional if and only if its monad is the cyclic monadic hull of the set of its essential points.*

As corollaries we cite the following Boolean-valued analogs of some classical criteria of A. Robinson.

8.5. Theorem. (1) *A standard set is the lowering of a compact space if and only if each of its essential points is near-standard.*

(2) *A standard set is the lowering of a totally bounded space if and only if each of its essential points is pre-near-standard.*

References

1. G. P. Akilov, E. V. Kolesnikov, and A. G. Kusraev, *On the order-continuous extension of a positive operator*, Sibirsk. Mat. Zh. **29** (1988), no. 5, 24–35; English transl. in Siberian Math. J. **29** (1988), no. 5.

2. K. I. Beidar and A. V. Mikhalev, *Orthogonal completeness and algebraic systems*, Uspekhi Mat. Nauk **40** (1985), no. 6, 79–115; English transl. in Russian Math. Surveys **40** (1985), no. 6.

3. P. Vopenka, *Mathematics in the Alternative Set Theory*, Teubner, Leipzig, 1979.

4. B. Z. Vulikh, *Introduction to the theory of partially ordered spaces*, Fizmatgiz, Moscow, 1961; English transl. Noordhoof, Groningen, 1967.

5. R. Goldblatt, *Topoi: The Categorial Analysis of Logic*, North-Holland, Amsterdam, 1979.

6. E. I. Gordon, *Real numbers in Boolean-valued models of set theory and K-spaces*, Dokl. Akad. Nauk SSSR **237** (1977), no. 4, 773–775; English transl. in Soviet Math. Dokl. **18** (1977).

7. ——, *K-spaces in Boolean-valued models of the theory of sets*, Dokl. Akad. Nauk SSSR **258** (1981), no. 4, 777–780; English transl. in Soviet Math. Dokl. **23** (1981).

8. ——, *On theorems on the preservation of relations in K-spaces*, Sibirsk. Mat. Zh. **23** (1982), no. 3, 55–65; English transl. in Siberian Math. J. **23** (1982), no. 3 (1983).

9. ——, *Nonstandard finite-dimensional analogs of operators in* $L_2(R^n)$, Sibirsk. Mat. Zh. **29** (1988), no. 2, 45–59; English transl. in Siberian Math. J. **29** (1988), no. 2.

10. ——, *Relatively standard elements in E. Nelson's internal set theory*, Sibirsk Mat. Zh. **30** (1989), no. 1, 89–95; English transl. in Siberian Math. J. **30** (1989), no. 1.

11. E. I. Gordon and V. A. Lyubetskiĭ, *Some applications of nonstandard analysis in the theory of Boolean-valued measures*, Dokl. Akad. Nauk SSSR **256** (1981), no. 5, 1037–1041; English transl. in Soviet Math. Dokl. **23** (1981).

12. A. K. Zvonkin and M. A. Shubin, *Nonstandard analysis and singular perturbations of ordinary differential equations*, Uspekhi Mat. Nauk **39** (1984), no. 2, 77–127; English transl. in Russian Math. Surveys **39** (1984), no. 2.

13. M. Davis, *Applied Nonstandard Analysis*, Wiley-Interscience, New York, 1977.

14. T. Jech, *Lectures in Set Theory with Particular Emphasis on Forcing*, Springer, Berlin, 1971.

15. V. G. Kanoveĭ, *Correctness of the Euler method of decomposing the sine function into an infinite product*, Uspekhi Mat. Nauk **43** (1988), no. 4, 57–81; English transl. in Russian Math. Surveys **43** (1988), no. 4.

16. L. V. Kantorovich, *Toward a general theory of operations in semiordered spaces*, Dokl. Akad. Nauk SSSR **1** (1936), no. 7, 271–274. (Russian)

17. ——, *On a class of functional equations*, Dokl. Akad. Nauk SSSR **4** (1936), no. 45, 211–216. (Russian)

18. L. V. Kantorovich and G. P. Akilov, *Functional Analysis*, "Nauka", Moscow, 1977; English transl., Pergamon Press, Oxford-New York, 1982.

19. L. V. Kantorovich, B. Z. Vulikh, and A. G. Pinsker, *Functional analysis in partially ordered spaces*, GITTL, Moscow, 1950. (Russian)

20. P. J. Cohen, *Set Theory and the Continuum Hypothesis*, Benjamin, New York, 1966.

21. A. G. Kusraev, *Some categories and functors of Boolean-valued analysis*, Dokl. Akad. Nauk SSSR **271** (1983), no. 1, 281–284; English transl. in Soviet Math. Dokl. **28** (1983).

22. ——, *Order-continuous functionals in Boolean-valued models of set theory*, Sibirsk Mat. Zh. **25** (1984), no. 1, 69–79; English transl. in Siberian Math. J. **25** (1984), no. 1.

23. ——, *Banach-Kantorovich spaces*, Sibirsk Mat. Zh. **26** (1985), no. 2, 119–126; English transl. in Siberian Math. J. **26** (1985), no. 2.

24. ——, *Vector Duality and its Applications*, "Nauka", Novosibirsk, 1985. (Russian)

25. ——, *Boolean-valued models and ordered algebraic systems*, in: Eighth All-Union Conference on Mathematical Logic. Abstracts of Lectures, Moscow, 1986, 99. (Russian)

26. ——, *Linear operators in lattice ordered spaces*, in: Studies in Geometry in the Large and Mathematical Analysis, "Nauka", Novosibirsk, 1987, 84–123. (Russian)

27. A. G. Kusraev and S. S. Kutateladze, *Analysis of subdifferentials by means of Boolean valued models*, Dokl. Akad. Nauk, SSSR **265** (1982), no. 5, 1061–1064; English transl. in Soviet Math. Dokl. **26** (1982/1983).

28. ——, *Subdifferential Calculus*, "Nauka", Novosibirsk, 1987. (Russian)

29. A. G. Kusraev and S. A. Malyugin, *Atomic decomposition of vector measures*, Sibirsk Mat. Zh. **30** (1989), no. 5, 101–110; English transl. in Siberian Math. J. **30** (1989), no. 5.

31. A. G. Kusraev and V. Z. Strizhevskiĭ, *Lattice normed spaces and majorized operators*, Studies in Geometry and Analysis, "Nauka", Novosibirsk, 1986, 56–102. (Russian)

32. S. S. Kutateladze, *Descents and Ascents*, Dokl. Akad. Nauk SSSR **272** (1983), no. 3, 521–524; English transl. in Soviet Math. Dokl. **28** (1983).

33. ——, *Infinitesimal tangent cones*, Sibirsk. Mat. Zh. **26** (1985), no. 6, 67–76; English transl. in Siberian Math. J. **26** (1985), no. 6.

34. ——, *Nonstandard methods in subdifferential calculus*, Partial Differential Equations, "Nauka", Novosibirsk, 1986, pp. 116–120. (Russian)

35. ——, *Cyclic monads and their applications*, Sibirsk Mat. Zh. **27** (1986), no. 1, 100–110; English transl. in Siberian Math. J. **27** (1986), no. 1.

36. ——, *A variant of nonstandard convex programming*, Sibirsk Mat. Zh. **27** (1986), no. 4, 84–92; English transl. in Siberian Math. J. **27** (1986), no. 4.

37. ——, *Directions of nonstandard analysis*, Trudy Inst. Mat. (Novosibirsk) **14** (1989), 153–182. (Russian)

38. ——, *Monads of pro-ultrafilters and of extensional filters*, Sibirsk Mat. Zh. **30** (1989), no. 1, 129–132; English transl. in Siberian Math. J. **30** (1989), no. 1.

39. ——, *Components of positive operators*, Sibirsk Mat. Zh. **30** (1989), no. 5, 111–119; English transl. in Siberian Math. J. **30** (1989), no. 5.

40. V. A. Lyubetskiĭ and E. I. Gordon, *Boolean extensions of uniform structures*, Studies in Nonclassical Logics and Formal Systems, "Nauka", Moscow, 1983, pp. 82–153. (Russian)

41. V. A. Lyubetskiĭ, *On some algebraic questions of nonstandard analysis*, Dokl. Akad. Nauk SSSR **280** (1985), no. 1, 38–41; English transl. in Soviet Math. Dokl. **31** (1985).

42. S. Albeverio, R. Hoegh-Krohn, J. E. Fenstad, and T. Lindstrom, *Nonstandard Methods in Stochastic Analysis and Mathematical Physics*, Academic Press, Orlando, FL, 1986.

43. W. A. J. Luxemburg (ed.), *Applications of Model Theory to Algebra, Analysis and Probability*, Holt, Rinehart and Winston, New York, 1969.

44. J. L. Bell, *Boolean Valued Models and Independence Proofs in Set Theory*, Clarendon Press, Oxford, 1979.

45. B. Benninghofen and M. M. Richter, *A general theory of superinfinitesimals*, Fund. Math. **128** (1987), no. 3, 199–215.

46. C. Cristiant, *Der Beitrag Gödels für die Rechfertigung der Leibnizschen Idee von der Infinitesimalen*, Osterreich. Akad. Wiss. Math. Nachr. **192** (1983), nos. 1–3, 25–43.

47. W. Henson and J. Keisler, *On the strength of nonstandard analysis*, J. Symbolic Logic **51** (1986), no. 2, 377–386.

48. R. B. Holmes, *Geometric Functional Analysis and its Applications*, Springer, Berlin-New York, 1975.

49. K. Hrbacek, *Nonstandard set theory*, Amer. Math. Monthly **86** (1979), no. 8, 659–677.

50. T. Jech, *Abstract theory of abelian operator algebras: An application of forcing*, Trans. Amer. Math. Soc. **289** (1985), no. 1, 133–162.

51. ——, *First order theory of complete Stonean algebras*, Canad. Math. Bull. **30** (1987), no. 4, 385–392.

52. I. Kaplansky, *Modules over operator algebras*, Amer. J. Math. **75** (1953), no. 4, 839–858.

53. T. Kawai, *Axiom systems of nonstandard set theory*, Logic Symposia, Hakone, 1979, 1980, Springer, Berlin-New York, 1981, pp. 57–65.

54. A. G. Kusraev, *Boolean valued convex analysis*, Mathematische Optimierung. Theorie und Anwendungen, Wartburg/Eisenach, 1983, pp. 106–109.

55. W. A. J. Luxemburg and A. C. Zaanen, *Riesz Spaces*, vol. 1, North-Holland, Amsterdam-London, 1971.

56. D. Maharam, *On kernel representation of linear operators*, Trans. Amer. Math. Soc. **70** (1955), no. 1, 229–255.

57. E. Nelson, *Radically Elementary Probability Theory*, Princeton University Press, Princeton, N.J., 1987.

58. ——, *The syntax of nonstandard analysis*, Ann. Pure Appl. Logic **38** (1988), no. 2, 123–134.

59. M. Ozawa, *A classification of type* I AW^**-algebras and Boolean valued analysis*, J. Math. Soc. Japan **36** (1984), no. 4, 589–608.

60. ——, *A transfer principle from von Neumann algebras to* AW^**-algebras*, J. London Math. Soc. (2) **32** (1985), no. 1, 141–148.

61. ——, *Boolean valued approach to the trace problem of* AW^**-algebras*, J. London Math. Soc. (2) **33** (1986), no. 2, 347–354.

62. ——, *Embeddable* AW^**-algebras and regular completions*, J. London Math. Soc. (2) **34** (1986), no. 3, 511–523.

63. A. Robinson, *Non-standard Analysis*, North-Holland, Amsterdam-London, 1970.

64. R. Solovay and S. Tennenbaum, *Iterated Cohen extensions and Souslin's problem*, Ann. of Math. **94** (1972), no. 2, 201–245.

65. G. Takeuti, *Two Applications of Logic to Mathematics*, Ivanami & Princeton University Press, Tokyo-Princeton, N.J., 1978.

66. ——, *Von Neumann algebras and Boolean valued analysis*, J. Math. Soc. Japan **35** (1983), no. 1, 1–21.

67. ——, C^**-algebras and Boolean valued analysis*, Japan J. Math. **9** (1983), no. 2, 207–246.

68. G. Takeuti and W. M. Zaring, *Axiomatic Set Theory*, Springer, New York, 1973.
69. D. Tall, *The calculus of Leibniz—an alternative modern approach*, Math. Intelligencer **2** (1979/80), no. 1, 54.
70. A. Z. Zaanen, *Riesz Spaces*, Vol. II, North-Holland, Amsterdam, 1983.

Translated by D. LOUVISH

Amer. Math. Soc. Transl.
(2) Vol. **151**, 1992

Automorphism Groups of Three-Dimensional Hyperbolic Manifolds

UDC 512.44.43: 515.165

A. D. MEDNYKH

In the last decade, thanks to the work of W. Thurston, his students and disciples, there has been growing interest in the study of three-dimensional hyperbolic manifolds that can be defined as a quotient-space H^3/Γ, where Γ is a discrete isometry group of the Lobachevskiĭ space acting without fixed points.

The theory of three-dimensional hyperbolic manifolds is similar in many respects to the theory of hyperbolic Riemann surfaces but, as the results given below show, in some cases the theories differ sharply from each other.

The subject of our discussion is automorphism groups of hyperbolic manifolds H^n/Γ, $n = 2$ or 3, that have finite hyperbolic volume. An automorphism of H^n/Γ is defined as an isometry of it that preserves the metric induced from H^n. Two-dimensional hyperbolic manifolds will by tradition be called Riemann surfaces, and isometries of them that preserve the orientation will be called conformal automorphisms.

As a rule, each of the sections below will start with well-known results from the theory of Riemann surfaces, and conclude with the corresponding results obtained for three-dimensional manifolds.

§1. General facts from the theory of automorphism groups of hyperbolic manifolds

Let $S = H^2/\Gamma$ be a compact Riemann surface of genus $g > 1$. Then, as is well known (H. A. Schwarz) the automorphism group $\operatorname{Aut} S$ of the surface S is finite. A similar result holds for a Riemann surface that has finite hyperbolic area. The group $\operatorname{Out}\Gamma$ of outer automorphisms of Γ is canonically realized as the group of biholomorphic automorphisms of the Teichmüller

1991 *Mathematics Subject Classification.* Primary 20F32, 51M10, 57S30; Secondary 30F99.

space Tg of Riemann surfaces of genus g. This is a finite-dimensional complex space of dimension $3g - 3$ and $\operatorname{Out}\Gamma$ acts on it properly discontinuously. Hence, in particular, it follows that $\operatorname{Out}\Gamma$ is countable ([13a], p. 228).

The group $\operatorname{Aut} S$ is naturally isomorphic to the stabilizer of the group $\operatorname{Out}\Gamma$ at the point of Tg corresponding to the complex structure of the surface (generally speaking, there are countably many such points), and so it is embedded in $\operatorname{Out}\Gamma$ as a finite subgroup (Harvey [24]).

Conversely, from a positive solution of the Fenchel-Nielsen-Teichmüller realization problem it follows that every finite subgroup of $\operatorname{Out}\Gamma$ is realized as the automorphism group $\operatorname{Aut} S$ of some Riemann surface S whose fundamental group is isomorphic to Γ (Kerckhoff [26]).

Baily [17] was the first to give a complete proof of the following conjecture, due to A. Hurwitz.

For any integer $g > 2$ there is a compact Riemann surface of genus g for which the group of all conformal automorphisms is trivial. Later Greenberg [22] showed that when $g > 2$ almost all points of the Teichmüller space Tg, except for an analytic subset, correspond to Riemann surfaces with a trivial conformal automorphism group. Despite this, not many constructive examples of such Riemann surfaces are known. One of them was constructed by Accola [14]. However, Accola's method does not enable us to describe analytically a fundamental set and generators of a group Γ that uniformizes the surface $S = H^2/\Gamma$. In [35] we gave an explicit construction of a group Γ for which the Riemann surface $S = H^2/\Gamma$ does not have nontrivial conformal automorphisms.

Unfortunately, the list of general results for three-dimensional hyperbolic manifolds is still not very long.

Let $M = H^3/\Gamma$ be a compact three-dimensional hyperbolic manifold. Thurston [46] observed that from Mostow's rigidity theorem [40] it follows that the group $\operatorname{Aut} M$ of automorphisms of M is always finite and canonically isomorphic to the group $\operatorname{Out}\Gamma$. By the rigidity theorem established by Prasad [41] a similar result holds for three-dimensional hyperbolic manifolds that have finite volume.

These rigidity theorems guarantee that the Teichmüller space of M consists of a single point, and so the analogues of most of the results given above do not make sense in the three-dimensional case.

We just observe that the existence of compact three-dimensional hyperbolic manifolds with trivial isometry group follows from a theorem of Kojima [27], formulated below in §4.

§2. Volumes of hyperbolic manifolds and orbifolds

Let Γ be a discrete group of isometries of the Lobachevskiĭ plane. The quotient space $\Omega^2 = H^2/\Gamma$, endowed with the hyperbolic metric induced from H^2, will be called a two-dimensional hyperbolic orbifold. In particular, when Γ acts on H^2 without fixed points, Ω^2 represents an ordinary hyperbolic Riemann surface. The three-dimensional hyperbolic orbifold $\Omega^3 = H^3/\Gamma$ is defined similarly. The concepts of hyperbolic area and hyperbolic volume in H^2 and H^3 are carried over naturally to Ω^2 and Ω^3.

Everywhere below, unless we say otherwise, we shall be dealing with orbifolds of finite area and finite volume.

The main instrument for studying automorphism groups of Riemann surfaces is the Riemann-Hurwitz formula. Let us write it in the language of the theory of orbifolds.

Let H^2/Γ_1 and H^2/Γ_2 be hyperbolic orbifolds of finite area such that $\Gamma_1 \subseteq \Gamma_2$. Then

$$\mathrm{Area}(H^2/\Gamma_1) : \mathrm{Area}(H^2/\Gamma_2) = |\Gamma_2 : \Gamma_1|, \tag{1}$$

where $|\Gamma_2 : \Gamma_1|$ is the index of the subgroup Γ_1 in Γ_2. (This is the Riemann-Hurwitz formula.)

Formula (1) is usually complemented by the following version of the Gauss-Bonnet formula:

$$(H^2/\Gamma) = -2\pi\chi(H^2/\Gamma), \tag{2}$$

where $\chi(H^2/\Gamma)$ is the Euler characteristic of the orbifold H^2/Γ; the definition and simple ways of calculating this can be found in [43]. In the particular case when Γ acts on H^2 without fixed points, and $S = H^2/\Gamma$ is a compact Riemann surface, $\chi(S)$ is the same as the Euler characteristic of S, understood in the usual sense. In this case $\chi(S)$ is a negative integer, and $\mathrm{Area}(S)$ is a positive number from the set $2\pi\mathbb{N}$, where $\mathbb{N}$ is the set of natural numbers. Hence we obtain the following result.

THEOREM 1. *The set of areas of hyperbolic surfaces is a discrete subset on the number line of the form* $2\pi\mathbb{N}$.

We observe that the least area in Theorem 1 is 2π. Exactly four nonhomeomorphic hyperbolic surfaces have this area: a sphere with three handles, a torus with one handle, a compact nonorientable surface of genus three, and a nonorientable surface of genus two with one handle.

The arguments given above enable us to establish the following result.

THEOREM 1°. *The set of areas of orientable hyperbolic surfaces is a discrete subset on the number line of the form* $2\pi\mathbb{N}$.

An orbifold $\Omega^n = H^n/\Gamma$ ($n = 2$ or 3) is said to be orientable if all the transformations of the group Γ preserve the orientation in H^n.

The Riemann-Hurwitz formula (1) carries over directly to the three-dimensional case. Namely, if H^3/Γ_1 and H^3/Γ_2 are hyperbolic orbifolds of finite volume such that $\Gamma_1 \subseteq \Gamma_2$, then

$$\mathrm{Volume}(H^3/\Gamma_1) : \mathrm{Volume}(H^3/\Gamma_2) = |\Gamma_2 : \Gamma_1|, \tag{3}$$

where $|\Gamma_2 : \Gamma_1|$ is the index of the subgroup Γ_1 in the group Γ_2.

However, the Gauss-Bonnet formula (2) for three-dimensional orbifolds ceases to be meaningful. The fact is that in the most interesting cases (for example, when $\Omega^3 = H^3/\Gamma$ is a compact three-dimensional hyperbolic manifold) we have $\chi(\Omega^3) = 0$ ([6]). In addition, as the results of [1], p. 135 show, there are no convenient formulas for calculating the hyperbolic volume of H^3/Γ even in the simplest cases.

In the three-dimensional situation, instead of Theorem 1 we have the following remarkable theorem of Thurston and Jørgensen [46].

THEOREM 2. *The set of volumes of three-dimensional hyperbolic manifolds forms a well-ordered subset of the number line of type* ω^ω.

In particular, it follows from Theorem 2 that there is a three-dimensional hyperbolic manifold of least volume. Meyerhoff [38] apparently obtained a far from exact estimate for its volume. Adams [15] discovered a noncompact hyperbolic manifold of least volume. It turned out to be the nonorientable manifold constructed in 1912 by Gieseking [21]. It was obtained by identifying the faces of a regular ideal hyperbolic tetrahedron in pairs. We observe that the complement to a "figure eight" in a three-dimensional sphere serves as a double orientable covering of the Gieseking manifold ([39]).

As in the two-dimensional case, an analogue of Theorem 2 holds for orientable manifolds.

THEOREM 2°. *The volumes of three-dimensional orientable hyperbolic manifolds form a well-ordered subset of the number line of type* ω^ω.

An orientable hyperbolic manifold that aspires to the least volume is known ([47], Ch. 4). It is obtained by generalized Dehn surgery (5.1), (5.2) on the complement to the Whitehead link in the three-dimensional sphere and has volume approximately equal to 0.94272 (see also [19]). We note that this manifold was constructed independently and by different methods in [8], where sufficiently weighty arguments were also given in favour of its minimality.

Analogues of Theorems 2 and 2° also hold for three-dimensional orbifolds ([18]).

We restrict ourselves to a statement of the theorem for the orientable case.

THEOREM 3. *The set of volumes of three-dimensional orientable manifolds forms a well-ordered subset of the number line of type* ω^ω.

It is surprising that for so long nobody observed that an analogue of the Thurston-Jørgensen theorem "in embryo" holds for two-dimensional hyperbolic orbifolds. Namely, as Mirkina showed [12], the following theorem holds.

THEOREM 4. *The set of volumes of two-dimensional orientable hyperbolic orbifolds forms a well-ordered subset of the number line of type* ω^ω.

A similar theorem can be stated for all (orientable or not) two-dimensional hyperbolic orbifolds.

Orbifolds that are smallest in area have been known from the time of Hurwitz. The least area of an orbifold H^2/Γ is equal to $\pi/42$ and is attained for the group Γ generated by reflections in the sides of a non-Euclidean triangle with angles $\pi/2$, $\pi/3$, and $\pi/7$. Among orientable orbifolds the least area is $\pi/21$ and is attained for H^2/Γ^+, where Γ^+ is the subgroup of Γ consisting of all transformations that preserve the orientation.

In the three-dimensional case only a noncompact orientable orbifold that is least in volume is known [37]. Its volume is $1/24$ of the volume of a regular ideal tetrahedron in H^3.

§3. The order of the automorphism group of a hyperbolic manifold

A classical theorem of Hurwitz asserts that the order of the group of conformal automorphisms of a compact Riemann surface of genus g does not exceed $84(g-1)$ ([25]). This estimate is exact, and is attained for the Klein surface of genus $g = 3$, and also for infinitely many other values of g ([32]).

The idea of proof of Hurwitz' theorem in the language of the theory of orbifolds is exceptionally simple.

Let $S = H^2/\Gamma$ be a compact Riemann surface of genus g, and let G be the group of conformal automorphisms of S. Let Γ_0 denote the lifting of G to the universal covering of H^2. By §2 we have

$$\mathrm{Area}(H^2/\Gamma) = 4\pi(g-1). \tag{4}$$

Moreover, since $G = \Gamma_0/\Gamma$, from (1) for the order $|G|$ of the group G we obtain

$$|G| = |\Gamma_0 : \Gamma| = \mathrm{Area}(H^2/\Gamma) : \mathrm{Area}(H^2/\Gamma_0). \tag{5}$$

Hence, observing from §2 that $\mathrm{Area}(H^2/\Gamma_0) \geq \pi/21$, and taking account of (4), we have

$$|G| \leq 4\pi(g-1)/(\pi/21) = 84(g-1). \tag{6}$$

In exactly the same way, for the complete group of automorphisms of S, those that do or do not preserve the orientation, we obtain

$$|\mathrm{Aut}\, S| \leq 4\pi(g-1)/(\pi/42) = 168(g-1). \tag{7}$$

Singerman [45] showed that the estimate (6), like the estimate (7), is attained for the Klein surface of genus 3, and for infinitely many other values of g.

The arguments given above carry over easily to the three-dimensional case. Namely, let $G = \mathrm{Aut}\, M$ be the automorphism group of the hyperbolic manifold $M = H^3/\Gamma$, and let Γ_0 be the lifting of G to the universal covering of H^3. Let $V_{\min}$ denote the least volume of a three-dimensional hyperbolic orbifold. From (3) we have

$$\begin{aligned}|\mathrm{Aut}\, M| &= |\Gamma_0 : \Gamma| = \mathrm{Volume}(H^3/\Gamma): \\ (H^3/\Gamma_0) &\leq \mathrm{Volume}(M)/V_{\min}.\end{aligned} \tag{8}$$

For noncompact hyperbolic manifolds of finite volume the estimate (8) could be improved. In fact, by [37] the volume of a noncompact (orientable or not) orbifold H^3/Γ_0 is bounded below by a value equal to $1/24$ of the volume of a regular ideal tetrahedron T in H^3, all four vertices of which lie on the absolute. As a result, making (8) more precise, we have

$$|\mathrm{Aut}\, M| \leq 24\frac{\mathrm{Volume}(M)}{\mathrm{Volume}(T)}. \tag{9}$$

We note [39] that $\mathrm{Volume}(T) = 1.01494\ldots$. Using a theorem of Mal′tsev [7] on the residual finiteness of finitely generated matrix groups, we can show that equality in (9) is attained for infinitely many nonhomeomorphic manifolds.

§4. Realization of a finite group as the automorphism group of a compact hyperbolic manifold

Greenberg [23] proved the following theorem. Let G be an arbitrary nontrivial finite group, and S an arbitrary compact Riemann surface. Then there is a regular branched covering $\pi: T \to S$ whose covering transformation group is isomorphic to G and coincides with the complete group of conformal automorphisms of the Riemann surface T. From this result we obtain two important consequences.

THEOREM 5. *Every finite group can be realized as the full group of conformal automorphisms of some compact Riemann surface.*

THEOREM 6. *Every field of algebraic functions of one complex variable has a Galois extension with any preassigned finite group.*

The complete analogue of Greenberg's theorem for three-dimensional hyperbolic manifolds is not known at present. However, Kojima [27] recently established the following result.

THEOREM 7. *Every finite group can be realized as the complete isometry group of some compact three-dimensional hyperbolic manifold.*

The following interesting question is still open: what is the analogue of Theorem 7 in the language of the theory of functions?

§5. Construction of three-dimensional hyperbolic manifolds

Answering Koebe's question about the existence of three-dimensional hyperbolic forms in the affirmative, Löbell [31] constructed the first example of a compact orientable manifold. Two years later there appeared the elegant example of the hyperbolic space of a dodecahedron of Seifert and Weber [44]. Al-Jubori [16] constructed the first example of a compact nonorientable hyperbolic manifold.

The paper of Löbell, based on complicated geometrical constructions, remained unnoticed for a long time.

Below we shall consider a general construction that enables us to construct the Löbell and Al-Jubori manifolds from a unified viewpoint.

Let P be a polyhedron in H^3, all of whose dihedral angles are right angles. Necessary and sufficient conditions for the existence of such polyhedra are contained in a paper of Andreev [2]. Let $\Delta(P)$ denote the group generated by reflections in the faces of P. Let $\mathbb{Z}_2 \oplus \mathbb{Z}_2 \oplus \mathbb{Z}_2$ be the direct sum of three cyclic groups of order 2, in which we distinguish the elements $a = (1, 0, 0)$, $b = (0, 1, 0)$, $c = (0, 0, 1)$, and $d = (1, 1, 1)$.

Let $\varphi: \Delta(P) \to \mathbb{Z}_2 \oplus \mathbb{Z}_2 \oplus \mathbb{Z}_2$ be a group homomorphism whose values on the reflections in the faces of P (the generators of the group $\Delta(P)$) belong to the set $\{a, b, c, d\}$. We observe that φ colors the faces of P in the four colors a, b, c, d. In fact, to each face Σ of P there corresponds uniquely an element $\varphi(\sigma)$ of the set $\{a, b, c, d\}$, where σ is the reflection in the face Σ. Conversely, from each coloring of the faces of P by the elements a, b, c, d we can uniquely recover the homomorphism φ.

As usual, we call a coloring of the faces of P regular if any two adjacent faces are colored in different colors. Vesnin [3] established the following curious result.

THEOREM 9. *To each regular coloring of the faces of P in four colors there corresponds an epimorphism $\varphi: \Delta(P) \to \mathbb{Z}_2 \oplus \mathbb{Z}_2 \oplus \mathbb{Z}_2$ such that the group $\Gamma = \operatorname{Ker}\varphi$ acts on H^3 without fixed points and $M^3 = H^3/\Gamma$ is a compact orientable hyperbolic manifold obtained by gluing together eight copies of P.*

Conversely, if Γ is a normal subgroup of index 8 in the group $\Delta(P)$ that acts on H^3 without fixed points and is such that H^3/Γ is an orientable manifold, then the canonical epimorphism

$$\varphi: \Delta(P) \to \Delta(P) \cong \mathbb{Z}_2 \oplus \mathbb{Z}_2 \oplus \mathbb{Z}_2$$

determines a regular coloring of the faces of P by four elements of the group $\mathbb{Z}_2 \oplus \mathbb{Z}_2 \oplus \mathbb{Z}_2$.

This theorem admits the following generalization.

Let v be one of the vertices of P, and $\Delta(P)|_v$ the stabilizer of the group $\Delta(P)$ at the point v. We note that the group $\Delta(P)|_v$ is generated by the reflections in the three mutually perpendicular faces of P passing through v, and is isomorphic to $\mathbb{Z}_2 \oplus \mathbb{Z}_2 \oplus \mathbb{Z}_2$.

THEOREM 10. *Let $\varphi: \Delta(P) \to \mathbb{Z}_2 \oplus \mathbb{Z}_2 \oplus \mathbb{Z}_2$ be an epimorphism having the following properties:*

a) *for any vertex* v *of the polyhedron* P *the restriction of* φ *to the group* $\Delta(P)|_v$ *is an isomorphism*;

b) φ *determines a regular coloring of* P *by at least five colors.*

Then $\Gamma = \operatorname{Ker}\varphi$ *acts on* H^3 *without fixed points, and* H^3/Γ *is a compact nonorientable hyperbolic manifold.*

The proof of the first assertion of the theorem is based on the argument that every nontrivial element x of the group $\Delta(P)$ that has fixed points in H^3 is conjugate to one of the elements $\tilde{x}$ of the group $\Delta(P)|_v$ for a suitable v. We observe that $\varphi(x) = \varphi(\tilde{x})$ and $\tilde{x}$ is a nontrivial element of the group $\Delta(P)|_v$. Hence by condition a) we have $\varphi(x) \neq 1$. Consequently, $x \notin \Gamma = \operatorname{Ker}\varphi$.

To prove the second assertion we consider the reflections $\sigma_1, \dots, \sigma_5$ in five differently colored faces of P. In this case $\varphi(\sigma_1), \dots, \varphi(\sigma_5)$ are distinct nontrivial elements of the group $\mathbb{Z}_2 \oplus \mathbb{Z}_2 \oplus \mathbb{Z}_2$ and so among them there are three such that

$$\varphi(\sigma_{i_1})\varphi(\sigma_{i_2})\varphi(\sigma_{i_3}) = \varphi(\sigma_{i_1}\sigma_{i_2}\sigma_{i_3}) = 1.$$

Hence the element $\sigma_{i_1}\sigma_{i_2}\sigma_{i_3} \in \Gamma$ and does not preserve the orientation of the space H^3.

Theorems 9 and 10 easily enable us to construct the Löbell and Al-Jubori manifolds mentioned above. To this end we consider in H^3 a rectangular $(2n+2)$-hedron P_n whose upper and lower bases are regular n-gons, and the lateral surface is formed from $2n$ pentagons. By Andreev's theorem such polyhedra exist for all $n \geq 5$. We observe that the polyhedron P_6 was first constructed in [31], and P_5 is the regular rectangular dodecahedron considered in [16].

Coloring P_6 in four colors by the elements a, b, c, d, of the group $\mathbb{Z}_2 \oplus \mathbb{Z}_2 \oplus \mathbb{Z}_2$ in one of the three possible ways, by Theorem 9 we obtain the Löbell manifold [31]. Coloring P_5 in a suitable way by the seven nontrivial elements of the group $\mathbb{Z}_2 \oplus \mathbb{Z}_2 \oplus \mathbb{Z}_2$, by Theorem 10 we obtain the Al-Jubori manifold [16]. Vesnin [3] showed that for each $n \geq 5$ there are regular colorings on P_n corresponding to both orientable and nonorientable hyperbolic manifolds.

In particular, when $n = 6$ this bears out Löbell's remark [31] that his method of construction also makes it possible to construct a compact nonorientable three-dimensional hyperbolic manifold. So apparently the priority in the construction of these manifolds belongs to Löbell, and not to Al-Jubori.

§6. Hyperelliptic Riemann surfaces and three-dimensional manifolds

A hyperbolic manifold $M^n = H^n/\Gamma$ ($n = 2$ or 3) is said to be hyperelliptic if there is an isometric involution τ on it such that $M^n/\langle\tau\rangle$ is homeomorphic to an n-dimensional sphere S^n. In this case the transformation is called a hyperelliptic involution. Hyperelliptic Riemann surfaces are widely known in complex analysis ([20], Ch. 3). The following theorem [36] answers the question of the existence of hyperelliptic manifolds in the three-dimensional case.

THEOREM 11. *Let P be a polyhedron in H^3 whose dihedral angles are right angles. Let $\Delta(P)$ denote the group generated by reflections in the faces of P. Suppose that the vertices and edges of P form a Hamiltonian graph. Then $\Delta(P)$ has a subgroup Γ of index* 8 *that acts on H^3 without fixed points and is such that $M^3 = H^3/\Gamma$ is a hyperelliptic manifold.*

We can take as a polyhedron P satisfying the condition of the theorem a dodecahedron, a truncated icosahedron (like a modern soccer ball), or the generalized Löbell polyhedron $P = P_n$, $n \geq 5$, considered above.

The hyperelliptic manifold M^3 in Theorem 11 is constructed as follows. Consider a Hamiltonian cycle γ consisting of edges of P and passing through each vertex of P. Then γ splits the surface of P into two simply-connected regions. It is not difficult to see that each of these regions has a regular coloring in two colors. Let us color one of them by the elements a and b, and the other by the elements c and d from the group $\mathbb{Z}_2 \oplus \mathbb{Z}_2 \oplus \mathbb{Z}_2$. As a result there arises the epimorphism $\varphi: \Delta(P) \to \mathbb{Z}_2 \oplus \mathbb{Z}_2 \oplus \mathbb{Z}_2$, whose kernel $\Gamma = \operatorname{Ker} \varphi$, by Theorem 9, acts on H^3 without fixed points and is a compact orientable hyperbolic manifold $M^3 = H^3/\Gamma$.

The action of a hyperelliptic involution on a manifold can be described as follows. Let A and B be reflections in adjacent faces of P colored by a and b respectively. Then $\tau = \Gamma AB$ is a hyperelliptic involution on $M^3 = H^3/\Gamma$ that acts according to the rule $\Gamma AB(\Gamma x) = \Gamma(ABx)$, $x \in H^3$.

The transformation τ does not change if the reflections A and B are replaced by any other pair of reflections in adjacent faces of P that are not separated by the Hamiltonian cycle γ.

The following properties of a hyperelliptic Riemann surface S are well known:

a) S has a unique hyperelliptic involution;

b) every conformal automorphism of S can be lowered to a conformal automorphism of the two-dimensional sphere $S^2 \cong S/\langle\tau\rangle$;

c) the group of conformal automorphisms of S acts effectively on the first homology group $H_1(S)$.

As the next theorem shows, three-dimensional hyperelliptic manifolds do not have these properties, generally speaking ([11], p. 255).

THEOREM 12. *Let P be a regular right-angled dodecahedron in the Lobachevskiĭ space H^3, and $\Delta(P)$ the group generated by reflections in the faces of P. Then there is a normal subgroup Γ of index* 8 *in $\Delta(P)$ that acts on H^3 without fixed points and is such that $M^3 = H^3/\Gamma$ is a hyperelliptic manifold with the following properties:*

(i) *M^3 has three hyperelliptic involutions τ_0, τ_1, τ_2;*

(ii) *there is an isometry* h *on* M^3 *such that*

$$h\tau_i h^{-1} = \tau_{i+1}, \qquad i = 0, 1, 2 \pmod 3.$$

In particular, h *cannot be lowered to a homeomorphism of the three-dimensional sphere* $S^3 \cong M^3/\langle\tau_0\rangle$;

(iii) τ_0, τ_1, *and* τ_2 *act trivially on the first homology group* $H_1(M^3)$.

To prove the theorem we use the following fact, known since Hamilton's time: any two Hamiltonian cycles on a dodecahedron P go into each other by a suitable isometry of P. This fact enables us to observe that the hyperelliptic manifold $M^3 = H^3/\Gamma$, constructed in the previous theorem uniquely (up to an isometry) from the Hamiltonian cycle, is determined by P itself. Generally speaking, different Hamiltonian cycles on P induce different hyperelliptic involutions on M^3. Selecting an isometry of P that cyclically permutes the three Hamiltonian cycles, we see that its lifting to M^3 determines an isometry that cyclically links the three hyperelliptic involutions. We have thus established assertions (i) and (ii). Assertion (iii) can be verified by direct calculation.

§7. Automorphism groups of Riemann surfaces and three-dimensional hyperbolic manifolds

From the end of the last century to the present time innumerably many papers have been written on the theory of automorphism groups of Riemann surfaces. A fairly complete bibliography on this question can be found in [4] and [13].

Below we just dwell on some results from this theory whose analogues in the three-dimensional case are either already known or will probably be obtained in the near future.

Let S be a Riemann surface of finite hyperbolic area. Then there is a uniquely defined compact Riemann surface $\overline{S}$ from which S is obtained by "knocking out" finitely many points. By Riemann's theorem on removing singularities, every conformal automorphism of S can be extended to a conformal automorphism of $\overline{S}$. The automorphism group of S is embedded in the automorphism group of $\overline{S}$. By a theorem of Marden [34], every three-dimensional hyperbolic manifold M of finite volume is obtained from some compact manifold $\overline{M}$ by "throwing out" finitely many nonintersecting surfaces. In this case $\overline{M}$ is far from uniquely determined and depends on Dehn surgery carried out on each component of the boundary of M. As Kojima showed in the course of proving the fundamental theorem of [27], by suitable Dehn surgery $\overline{M}$ can be made hyperbolic, but generally speaking the isometry group of M is not embedded in the isometry group of $\overline{M}$.

Thus the study of conformal automorphisms of Riemann surfaces of finite hyperbolic area reduces to the study of conformal automorphisms of compact Riemann surfaces. For automorphism groups of three-dimensional hyperbolic manifolds of finite volume this approach cannot be used, unfortunately.

Conformal automorphism groups of hyperbolic surfaces of small genus, or equivalently of small hyperbolic area, have been completely described.

A Riemann surface of genus 2 is always hyperelliptic, and so it can be represented as a double-branched covering over a Riemann sphere. By virtue of the uniqueness of the hyperelliptic involution, the conformal automorphism group of the surface can be lowered to the conformal automorphism group of the sphere. Hence it can be effectively calculated ([20], Ch. 3).

A complete classification of conformal automorphism groups of a Riemann surface of genus 3 was carried out in [28]. The conformal automorphism groups of Riemann surfaces of genus 4 and 5 were described in [29] and [30].

Let us turn to a description of the automorphism groups of the "simplest" three-dimensional hyperbolic manifolds that have not too large hyperbolic volume.

From some old results of Magnus [33] and Mostow's rigidity theorem it follows that the complete isometry group of the complement of the "figure-eight" with respect to the three-dimensional sphere is isomorphic to the dihedral group of order 8.

Riley [42] calculated the isometry groups of the complements of Borromean rings, and also of some more complicated knots and links. The following results about isometry groups of compact hyperbolic manifolds were obtained in [9] and [11].

THEOREM 13. *The full isometry group of the Löbell manifold* [31] *is isomorphic to*

$$(\mathbb{Z}_2 \oplus \mathbb{Z}_2 \oplus \mathbb{Z}_2) \rtimes (D_6 \rtimes \mathbb{Z}_2)$$

THEOREM 14. *The full isometry group of the Al-Jubori manifold* [16] *is isomorphic to* $\mathbb{Z}_2 \oplus \mathbb{Z}_2 \oplus \mathbb{Z}_2$.

The following result was established in [10].

THEOREM 15. *The full isometry group of the hyperbolic space of the Seifert-Weber dodecahedron* [44] *acts faithfully on its first homology group.*

The main instrument for proving Theorems 13, 14, and 15 consists of the greatest lower bounds, obtained in [5], of the distances between fixed points of discrete groups acting in the plane and in Lobachevskiĭ space.

BIBLIOGRAPHY

1. D. B. Alekseevskiĭ, E. B. Vinberg, and A. S. Solodovnikov, *The geometry of spaces of constant curvature*, Modern Problems of Mathematics. Fundamental Directions, vol. 29, VINITI, Moscow, 1988, pp. 5–146. (Russian)

2. E. M. Andreev, *Convex polyhedra in Lobachevskiĭ space*, Mat. Sb. **81** (1970), 445–478; English transl. in Math. USSR-Sb. **10** (1970).

3. A. Yu. Vesnin, *Three-dimensional hyperbolic manifolds of Löbell type*, Sibirsk. Mat. Zh. **28** (1987), no. 5, 50–53; English transl. in Siberian Math. J. **28** (1987).

4. E. B. Vinberg and O. V. Shvartsman, *Riemann surfaces*, Algebra, Topology, Geometry, vol. 18, VINITI, Moscow, 1978, pp. 193–248. (Russian)

5. D. A. Derevnin and A. D. Mednykh, *Geometrical properties of discrete groups acting with fixed points in Lobachevskiĭ space*, Dokl. Akad. Nauk SSSR **300** (1988), 27–30; English transl. in Soviet Math. Dokl. **37** (1988).

6. H. Seifert and W. Threlfall, *Lehrbuch der Topologie*, Teubner, Leipzig, 1934 (reprinted by Chelsea Publ. Co., New York, 1947).

7. A. I. Mal′tsev, *On the faithful representation of infinite groups by matrices*, Mat. Sb. **8** (1940), 405–422; English transl. in Amer. Math. Soc. Transl. (2) **45** (1965), 1–18.

8. S. V. Matveev and A. T. Fomenko, *Constant energy surfaces of Hamiltonian systems, enumeration of three-dimensional manifolds in increasing order of complexity, and computation of volumes of closed hyperbolic manifolds*, Uspekhi Mat. Nauk **43** (1988), no. 1, 5–22; English transl. in Russian Math. Surveys **43** (1988), no. 1.

9. A. D. Mednykh, *Automorphism groups of three-dimensional hyperbolic manifolds*, Dokl. Akad. Nauk SSSR **285** (1985), 40–44; English transl. in Soviet Math. Dokl. **32** (1985).

10. ——, *On the isometry group of the hyperbolic space of the Seifert-Weber dodecahedron*, Sibirsk. Mat. Zh. **28** (1987), no. 5, 134–144; English transl. in Siberian Math. J. **28** (1987), no. 5.

11. ——, *Branched coverings and automorphism groups of hyperbolic manifolds*, Doctoral dissertation, Novosibirsk, 1988. (Russian)

12. M. V. Mirkina, *The two-dimensional analogue of Thurston's theorem on the volumes of hyperbolic orbifolds*, Proc. All-Union School on Optimal Control, Geometry and Analysis (29 September–9 October 1988), Kemerovo, 1988. (Russian)

13a. H. Zieschang, E. Vogt, and H. D. Coldeway, *Surfaces and planar discontinuous groups*, Lecture Notes in Math., vol. 835, Springer-Verlag, Berlin and New York, 1980.

13b. H. Zieschang, *Finite groups of mapping classes of surfaces*, Lecture Notes in Math., vol. 875, Springer-Verlag, Berlin and New York, 1981.

14. R. D. M. Accola, *Strongly branched coverings of closed Riemann surfaces*, Proc. Amer. Math. Soc. **226** (1970), 315–322.

15. C. Adams, *The noncompact hyperbolic* 3*-manifold of minimal volume*, Proc. Amer. Math. Soc. **100** (1987), 601–606.

16. N. K. Al-Jubori, *On non-orientable hyperbolic* 3*-manifolds*, Quart. J. Math. Oxford (2) **31** (1980), 9–18.

17. W. L. Baily, *On the automorphism group of a generic curve of genus* >2, J. Math. Kyoto Univ. **1** (1961/62), 101–108; correction p. 325.

18. A. Borel, *Commensurability classes and volumes of hyperbolic* 3*-manifolds*, Ann. Scuola Norm. Sup. Pisa Cl. Sci. (4) **8** (1981), 1–33.

19. T. Chinburg, *A small arithmetic hyperbolic three-manifold*, Proc. Amer. Math. Soc. **100** (1987), 140–144.

20. H. Farkas and I. Kra, *Riemann surfaces*, Graduate Texts in Math. 71, Springer-Verlag, New York and Berlin, 1980.

21. H. Gieseking, *Analytische Untersuchungen über topologische Gruppen*, Thesis, Münster, 1912.

22. L. Greenberg, *Maximal Fuchsian groups*, Bull. Amer. Math. Soc. **69** (1963), 569–573.

23. ——, *Maximal groups and signatures*, Ann. of Math. Studies No. 79, Princeton Univ. Press, Princeton, NJ, 1974, 207–226.

24. W. J. Harvey, *On branch loci in Teichmüller space*, Trans. Amer. Math. Soc. **153** (1971), 387–399.

25. A. Hurwitz, *Über algebraische Gebilde mit eindeutigen Transformationen in sich*, Math. Ann. **41** (1893), 403–442.

26. S. P. Kerckhoff, *The Nielsen realization problem*, Ann. of Math. (2) **117** (1983), 235–265.

27. S. Kojima, *Isometry transformations of hyperbolic* 3*-manifolds*, Topology Appl. **29** (1988), 297–307.

28. A. Kuribayashi and K. Komiya, *On the structure of the automorphism group of a compact Riemann surface of genus* 3, Bull. Fac. Sci. Engrg. Chuo Univ. Ser. I Math. **23** (1980), 1–34.

29. I. Kuribayashi and A. Kuribayashi, *On automorphism groups of compact Riemann surfaces of genus* 4, Proc. Japan Acad. Ser. A Math. Sci. **62** (1986), 65–72.

30. A. Kuribayashi and H. Kimura, *On automorphism groups of compact Riemann surfaces of genus* 5, Proc. Japan Acad. Ser. A Math. Sci. **63** (1987), 126–130.

31. F. Löbell, *Beispiele geschlossener dreidimensionaler Clifford-Kleinscher Räume negative Krümmung*, Ber. Sächs. Akad. Wiss. Leipzig **83** (1931), 167–174.

32. A. M. Macbeath, *On a theorem of Hurwitz*, Proc. Glasgow Math. Assoc. **5** (1961), 90–96.

33. W. Magnus, *Untersuchungen über einige unendliche discontinuierliche Gruppen*, Math. Ann. **105** (1931), 52–74.

34. A. Marden, *The geometry of finitely generated Kleinian groups*, Ann. of Math. (2) **99** (1974), 383–462.

35. A. D. Mednykh, *Hyperbolic Riemann surfaces with the trivial group of automorphisms*, Deformations of Mathematical Structures (J. Lawrynowicz, ed.), Kluwer Academic Publishers, Dordrecht, 1989, 115–125.

36. ——, *Three-dimensional hyperelliptic manifolds*, Ann. Global Anal. Geom. **8** (1990), no. 1, 13–19.

37. R. Meyerhoff, *The cusped hyperbolic* 3*-orbifold of minimal volume*, Bull. Amer. Math. Soc. (N.S.) **13** (1985), 154–156.

38. ——, *A lower bound for the volume of hyperbolic* 3*-manifolds*, Canad. J. Math. **39** (1987), 1038–1056.

39. J. Milnor, *Hyperbolic geometry: The first* 150 *years*, Bull. Amer. Math. Soc. (N.S.) **6** (1982), 9–24.

40. G. Mostow, *Quasi-conformal mappings in n-space and the rigidity of hyperbolic space forms*, Publ. Math. IHES **34** (1968), 53–104.

41. G. Prasad, *Strong rigidity of* **Q**-*rank* 1 *lattices*, Invent. Math. **21** (1973), 255–286.

42. R. Riley, *An elliptical path from parabolic representations to hyperbolic structures*, Topology of Low-dimensional Manifolds, Lecture Notes in Math. 722, Springer-Verlag, Berlin, 1979, 99–133.

43. P. Scott, *The geometries of* 3*-manifolds*, Bull. London Math. Soc. **15** (1983), 401–487.

44. H. Seifert and C. Weber, *Die beiden Dodekaederräume*, Math. Z. **37** (1933), 237–253.

45. D. Singerman, *Symmetries of Riemann surfaces with large automorphism groups*, Math. Ann. **210** (1974), 17–32.

46. W. Thurston, *The geometry and topology of* 3*-manifolds*, Princeton Univ. Lecture Notes, 1978.

47. J. Weeks, *Hyperbolic structures on three-manifolds*, Ph.D. thesis, Princeton Univ., 1985.

Translated by E. J. F. PRIMROSE

Amer. Math. Soc. Transl.
(2) Vol. **151**, 1992

Automorphic Forms, L-Functions, and p-adic Analysis

UDC 512.75

A. A. PANCHISHKIN

In this paper we discuss some recent results relating to the construction of p-adic L-functions associated with different classes of automorphic forms. This theory began with the work of Kubota and Leopoldt [21], who constructed a p-adic interpolation of the values of the Riemann zeta-function at the negative integers, and the work of Iwasawa [19], who applied the p-adic zeta-function to describe the class groups of cyclotomic fields. Since then, the class of functions having p-adic analogs has been continually growing. Yu. I. Manin [7], [9] and Mazur and Swinnerton-Dyer [2] created the theory of modular symbols, which enables one to construct nonarchimedean zeta-functions of automorphic forms on GL_2 over totally real fields [9] and fields of CM-type [6], [20]. The author's papers [11], [13]–[16] are devoted to the problem of constructing a general theory of nonarchimedean zeta-functions associated with automorphic forms on the classical groups. In particular, we examine in detail the case of the symplectic group of even genus. The general definition of the corresponding complex zeta-functions is due to Langlands [8], [12], and the nonarchimedean construction is equivalent to proving certain generalized Kummer congruences for special values of these zeta-functions. Here the Kummer congruences for the Bernoulli numbers correspond to the case of the Riemann zeta-function. We call attention to Manin's survey [9], which contains a list of unsolved problems in the theory of nonarchimedean zeta-functions, of which the problem we are considering is mentioned as the most complicated one (§9, section 9.9). We begin with a classical example, which leads to the construction of the Kubota-Leopoldt p-adic zeta-function.

1991 *Mathematics Subject Classification*. Primary 11S40, 11S80, 11F33; Secondary 11M41, 11F46, 11F85, 11F67.

§1. Kummer congruences and p-adic measures

It is well known that the Riemann zeta-function

$$\zeta(s)=\sum_{n=1}^{\infty} n^{-s} = \prod_{p\text{ prime}} (1-p^{-s})^{-1}$$

takes rational values at the nonnegative integers:

$$\zeta(1-k)=-B_k/k \quad (\text{for } k\geq 1), \tag{1}$$

where B_k is the kth Bernoulli number, defined by the expansion

$$\frac{te^t}{e^t-1}=\sum_{k=0}^{\infty}\frac{B_k}{k!}t^k=e^{Bt},$$

where we have used symbolic notation in which B_k is written instead of B^k. There are some well-known integrality properties of the numbers (1): for example, the Lipschitz-Sylvester theorem [10] says that for any positive integer $c>1$ the number $c^k(c^k-1)B_k/k$ is an integer (the proof follows from the power series expansion of the logarithmic derivative of the expression $(e^{ct}-1)/(e^t-1)$). In particular, if c is prime to a prime number p, then the following numbers are p-integral:

$$\zeta_{(p)}^{(c)}(-k)=(1-p^k)(1-c^{k+1})\zeta(-k) \quad \text{for } k\geq 0. \tag{2}$$

The classical Kummer congruences can be conveniently formulated in terms of the numbers (2): for any integral polynomial

$$h(X)=\sum a_k X^k \in Z[X],$$

the relation $h(x)\equiv 0 \pmod{p^N}$ for all $x\in\mathbb{Z}$ implies that

$$\sum a_k\zeta_{(p)}^{(c)}(-k)\equiv 0 \pmod{p^N}. \tag{3}$$

Kummer considered the polynomials $h(X)=X^{k'}-X^k$, where $k'\equiv k \bmod p^{N-1}(p-1)$; then one obtains $\zeta_{(p)}^{(c)}(-k')\equiv\zeta_{(p)}^{(c)}(-k) \bmod p^N$.

Later, the Kummer congruences (3) were explained in terms of the theory of p-adic integration. Let $\mathbb{Z}_p$ be the ring of p-adic integers, and let $\mathbb{Q}_p$ be the field of p-adic numbers, which is defined as the completion of the field of rational numbers with respect to the p-adic metric:

$$\left|p^n\frac{a}{b}\right|_p=p^{-n} \quad \text{for } (a,p)=(b,p)=1.$$

Here $\mathbb{Z}_p=\{x\in\mathbb{Q}_p \mid |x|_p\leq 1\}$, and the "$p$-small" numbers are those that are divisible by a large power of p. Now the congruences (3) are equivalent to the claim that there exists a finitely additive measure $\mu^{(c)}$ on $\mathbb{Z}_p$ with values in $\mathbb{Z}_p$ such that

$$\int_{\mathbb{Z}_p} x^k\,d\mu^{(c)}(x)=\zeta_{(p)}^{(c)}(-k) \quad (k\geq 0). \tag{4}$$

In fact, in order to obtain (3) from (4) one need only integrate the polynomial $h(X)$ and take into account that it takes small values, while the measure $\mu^{(c)}$ takes p-integral values. Conversely, to show that the congruences (3) imply the existence of the measure $\mu^{(c)}$, one must be able to determine the integral of any continuous function $\varphi\colon \mathbb{Z}_p \to \mathbb{Z}_p$:

$$\int_{\mathbb{Z}_p} \varphi(x)\, d\mu^{(c)}(x)$$

in such a way that the relations (4) hold, i.e., one must construct a measure given the values of the integral of monomials. One does this by approximating the continuous function φ by polynomials whose integral is determined from (4) and then taking the limit. It is the condition (3) that guarantees that the limit is well defined: the values of the integral of polynomials that take nearby values are also near to one another. The proof of the congruences (3) can be based upon the connection between the power sums $S_k(n) = \sum_{a=1}^{n-1} a^k$ and the Bernoulli polynomials

$$B_k(x) = (x+B)^k = \sum_{i=0}^{k} c_k^i B_i x^{k-i}.$$

The sums $S_k(n)$ can be expressed in terms of the Bernoulli polynomials as follows:

$$(k+1)S_k(n) = B_{k+1}(n) - B_{k+1}.$$

Then, in turn, the Bernoulli numbers can be recovered from the power sums using the limit formula

$$B_k = \lim_{N\to\infty} \frac{1}{pN} S_k(p^N) \quad (p\text{-adic limit}). \tag{5}$$

Given a polynomial $h(X)$, to verify (3) we now express the numbers

$$\zeta_{(p)}^{(c)}(-k)$$

in terms of Bernoulli numbers, and then approximate them by power sums using (5). The congruences (3) can easily be verified term-by-term: when each term is considered separately, the congruences become obvious because of the condition that the values of $h(X)$ are divisible by p^N.

§2. The p-adic zeta-function and the Mellin transform

The domain of definition of the classical zeta-functions is (the analytic Lie group)

$$X_\infty = \mathbb{C} = \operatorname{Hom}_{\text{contin}}(\mathbb{R}_+^\times, \mathbb{C}^\times) \qquad (y \mapsto y^s,\ s \in \mathbb{C}).$$

The Mellin transform is defined for a function φ on $\mathbb{R}_+^\times$ (which satisfies certain conditions on smoothness and vanishing at infinity) by the formula

$$L_\varphi(s) = \int_{\mathbb{R}_+^\times} \varphi(y) y^s \frac{dy}{y}, \quad L_\varphi\colon X_\infty \to \mathbb{C}.$$

In the p-adic theory one replaces $\mathbb{C}$ by the Tate field $\mathbb{C}_p = \overset{\Delta}{\mathbb{Q}}_p$ (the completion of the algebraic closure of the field of p-adic numbers) with norm $|\cdot|_p$ uniquely determined by the condition $|p|_p = p^{-1}$. This field is topologically complete and algebraically closed. The domain of definition of p-adic zeta-functions is the $\mathbb{C}_p$-analytic Lie group [5], [7]

$$X_p = X(\mathbb{Z}_p^\times) = \mathrm{Hom}_{\mathrm{contin}}(\mathbb{Z}_p^\times, \mathbb{C}_p^x) \tag{6}$$

where the symbol $X(\cdot)$ denotes the functor of p-adic characters of a topological group. The group X_p can be described as a disconnected union of discs

$$T = \{t \in \mathbb{C}_p \mid |t-1|_p < 1\} \subset \mathbb{C}_p.$$

In fact, let $\nu = 1$ or $\nu = 2$, depending on whether $p > 2$ or $p = 2$. Then $\mathbb{Z}_p^\times = (\mathbb{Z}/p^\nu\mathbb{Z})^\times \times \mathscr{U}$, where $\mathscr{U} = \langle 1 + p^\nu \rangle$, and we have the decomposition

$$X_p = X((\mathbb{Z}/p^\nu\mathbb{Z})^\times) \times X(\mathscr{U}) \tag{7}$$

where $\mathscr{U}$ is a closed topologically cyclic group isomorphic to $\mathbb{Z}_p$; the characters in the first factor are called tame characters, and the ones in the second factor are called wild characters, and we have the isomorphism

$$X(\mathscr{U}) \cong T, \quad \text{where } \chi \mapsto \chi(1+p^\nu) = t \in T. \tag{8}$$

By definition, an analytic function on X_p is given by a convergent power series on each component of the decomposition (7). If μ is a bounded measure on $\mathbb{Z}_p^\times$ with values in $\mathbb{C}_p$, then one can define its Mellin transform

$$L_\mu \colon X_p \to \mathbb{C}_p, \qquad \chi \mapsto \int_{\mathbb{Z}_p} \chi \, d\mu. \tag{9}$$

This transform gives an isomorphism between the algebra of bounded $\mathbb{C}_p$-valued measures on $\mathbb{Z}_p^\times$ with the convolution operation as its product and the algebra of bounded $\mathbb{C}_p$-analytic functions X_p with the operation of pointwise multiplication (the Iwasawa algebra). In each of the analytic components in (7), the restriction of this algebra can also be defined as $\mathscr{O}_p[[X]] \otimes \mathbb{Q}$, where $\mathscr{O}_p = \{x \in \mathbb{C}_p \mid |x_p| \le 1\}$. We have a Weierstrass preparation theorem for the elements of this algebra: the power series that occur can be written as the product of a certain polynomial and an invertible power series. In particular, any function in the Iwasawa algebra is uniquely determined once an infinite set of values is given on each analytic component in the decomposition (7). We shall be interested in the special values of analytic functions on the arithmetic characters—the elements of the group X_p. These characters include, in the first place, the characters of the form

$$x_p^k \colon y \mapsto y^k \qquad (k \in \mathbb{Z}, y \in \mathbb{Z}_p^\times),$$

and they also include the characters of finite order, which form the discrete subgroup X_p^{tors} of the group X_p. Among the elements $\chi \in X_p^{\mathrm{tors}}$ one finds all of the tame characters in the decomposition (7), and also the wild characters which correspond to the p-primary roots of unity in T under the isomorphism (8). We fix an imbedding $i_p: \overline{\mathbb{Q}} \to \mathbb{C}_p$ of the field of all algebraic numbers in the Tate field. Then any character of finite order $\chi: \mathbb{Z}_p^\times \to \mathbb{C}_p^x$ can be factored as the composition

$$\chi: \mathbb{Z}_p^\times \to (\mathbb{Z}/p^N)^\times \to \overline{\mathbb{Q}}^\times \to \mathbb{C}_p^\times$$

for some natural number N, which we shall choose to be the smallest possible for the given character χ. In this way we obtain a one-to-one correspondence between the elements $\chi \in X_p^{\text{tors}}$ and the Dirichlet characters of p-primary conductor, which we denote by the same letter χ; here c_χ will denote the conductor of the character χ, $c_\chi = p^N$ ($N = 0$ for the trivial χ).

We now return to the measure constructed above and consider its Mellin transform $L_{ju^{(c)}}: X_p \to \mathbb{C}_p$, which is a bounded $\mathbb{C}_p$-analytic function. We define the Kubota-Leopoldt p-adic zeta-function [21] by the equality

$$L_p(x) = (1 - cx(c))^{-1} L_{\mu^{(c)}}(x) \qquad (x \in X_p). \tag{10}$$

This function is $\mathbb{C}_p$-analytic on X_p, and it has a single simple pole at the point $x = x_p^{-1}$, corresponding to the pole of the Riemann zeta-function at $s = 1$. Moreover, for all $k \geq 0$ and $\chi \in X_p^{\text{tors}}$ (except for the case when $k = -1$ and χ is the trivial character) we have the relation

$$L_p(\chi x_p^k) = i_p[(1 - \chi(p)p^k)L(-k, \chi)], \tag{11}$$

where

$$L(s, \chi) = \sum_{n=1}^{\infty} \chi(n) n^{-s} = \prod_{q \text{ prime}} (1 - \chi(q) q^{-s})^{-1}$$

denotes the Dirichlet L-function with character χ; here both sides of (11) lie in the field $i_p(\mathbb{Q}(\chi))$, and the field $\mathbb{Q}(\chi)$ is the field obtained by adjoining to $\mathbb{Q}$ all of the roots of unity that are values of χ.

The equalities (11) are important in the study of the class numbers of the cyclotomic fields $\mathbb{Q}(\chi)$. The connection is given by the fundamental theorem of Iwasawa theory [19], which was recently proved in its entirety by Mazur and Wiles [23]. We further note that the special values (11) uniquely determine the function L_p, and hence the definition (10) does not depend upon the choice of the auxiliary number c. The function L_p was first constructed by Kubota and Leopoldt without p-adic integration using the theory of p-adic interpolation; the construction of the measure $\mu^{(c)}$ is due to Mazur [22]. Mazur's construction was subsequently generalized to other classes of zeta-functions that arise in number theory and the theory of automorphic forms (see the introduction). We shall give a detailed description of the extension of this construction to the case of the standard zeta-functions of Siegel modular forms, which were introduced and studied by A. N. Adrianov and V. L. Kalinin [4].

§3. Siegel modular forms

Let m be a natural number. We consider the algebraic group

$$G = \left\{ g = \begin{pmatrix} a & b \\ c & d \end{pmatrix} \in \mathrm{GL}_{2m} \middle| {}^t g J_m g = \mu(g) J_m, \ \mu(g) \in \mathrm{GL}_1 \right\},$$

where $J_m = \begin{pmatrix} 0 & -1_m \\ 1_m & 0 \end{pmatrix}$ is the standard symplectic matrix. The group

$$G_{\mathbb{R}}^+ = \{ g \in G_{\mathbb{R}} | \mu(g) > 0 \}$$

acts transitively by means of holomorphic automorphisms on the Siegel upper half-plane of genus m

$$H_m = \{ z \in \mathrm{GL}_m(\mathbb{C}) | {}^t z = z, \ \mathrm{Im}(z) > 0 \},$$

according to the formula $g(z) = (az+b)(cz+d)^{-1}$. The Siegel modular group $\Gamma_m = \mathrm{Sp}_m(\mathbb{Z})$ is a discrete subgroup of $G_{\mathbb{R}}^+$.

The complex vector space $\mathcal{M}_k^m$ of Siegel modular forms of genus m and weight k (where k is a natural number) consists of holomorphic functions $f\colon H_m \to \mathbb{C}$, which satisfy the automorphy condition

$$f(\gamma(z)) = \det(cz+d)^k f(z) \quad \text{for all } \gamma = \begin{pmatrix} a & b \\ c & d \end{pmatrix} \in \Gamma_m. \tag{12}$$

If $m = 1$, then one must also require that $|f(z)|$ be bounded as $\det \mathrm{Im}(z) \to \infty$, i.e., that the function be regular at $z = i\infty$; if $m > 1$, then this condition holds automatically, because of the "Koecher effect." Siegel modular forms can be expanded in a Fourier series of the form

$$f(z) = \sum_{\xi \geq 0} a(\xi) e_m(\xi z) \qquad (e_m(z) = \exp(2\pi i \, \mathrm{tr}(z))), \tag{13}$$

where the summation in (13) is taken over the set of all half-integral non-negative definite symmetric matrices (if $m > 1$, the condition that $\xi \geq 0$ follows from (12)). If the sum in (13) has only terms with positive definite $\xi > 0$, then f is called a cusp form. The cusp forms form a subspace $\mathcal{S}_k^m$ of the complex vector space $\mathcal{M}_k^m$. One similarly defines the spaces $\mathcal{U}_k^m(c, \psi)$ and $\mathcal{S}_k^m(c, \psi)$ of modular forms and cusp forms on H_m of weight k with Dirichlet character ψ mod c for the congruence subgroup

$$\Gamma_0^m(c) = \left\{ \gamma = \begin{pmatrix} a & b \\ c & d \end{pmatrix} \in \Gamma_m \middle| c \equiv 0 \bmod c \right\},$$

where c is a natural number.

One source of interesting examples of Siegel modular forms is the theta-series

$$\Theta_Q^t(z) = \sum_{X \in M_{n,m}(\mathbb{Z})} e_m(\mathbb{Q}[X]z)_{(Q[X] = {}^t XQX)}, \tag{14}$$

where Q is a positive definite even unimodular $n \times n$ matrix. In this case $\Theta_Q(z) \in \mathscr{U}^m_{n/2}$ and the integer n is divisible by 8. Another example is the Siegel-Eisenstein series

$$E^m_k(z) = \sum_{(c,d)} \det(cz+d)^{-k} \in \mathscr{U}^m_k \qquad (k > m+1), \tag{15}$$

where the summation in (15) is taken over the set of equivalence classes of symmetric pairs of relatively prime integral matrices; this means that

$$\{X \in M_m(\mathbb{Q}) | Xc, \ Xd \in M_m(\mathbb{Z})\} = M_m(\mathbb{Z}),$$

$c \cdot {}^t d = d \cdot {}^t c$ and $(c, d) \sim (c_1, d_1)$, if $c = uc_1$, $d = ud_1$ for some unimodular matrix $u \in \mathrm{GL}_m(\mathbb{Z})$. The set of such equivalence classes can be identified with the set of cosets of the form $(P \cap \Gamma_m)\left(\begin{smallmatrix} a & b \\ c & d \end{smallmatrix}\right) \in (P \cap \Gamma_m)\backslash\Gamma_m$, where $P = \{g = \left(\begin{smallmatrix} a & b \\ c & d \end{smallmatrix}\right) \in G_{\mathbb{R}}\}$ is the parabolic subgroup, and the condition $k > m+1$ is necessary for absolute convergence of the series (15).

§4. The Hecke algebra and zeta-functions

Before giving the general definition of the zeta-function of Siegel modular forms, we recall that when $m = 1$ the zeta-functions are defined using Fourier coefficients that satisfy certain multiplicativity conditions. The situation is more complicated when $m > 1$, and the zeta-functions of modular forms are introduced using the Hecke algebra associated to Siegel modular forms and its characters.

Let Γ be a subgroup of a semigroup Δ, where the pair (Γ, Δ) satisfies the following finiteness condition: for all $g \in \Delta$ the double coset $\Gamma g \Gamma$ is a disjoint union of finitely many left cosets

$$\Gamma g \Gamma = \bigcup_{c=1}^{\mathscr{H}(g)} \Gamma g_i. \tag{16}$$

Then the Hecke algebra $\mathscr{H} = \mathscr{H}(\Gamma, \Delta)$ of the pair (Γ, Δ) is defined to be the set of formal $\mathbb{C}$-linear combinations of double cosets $(\Gamma g \Gamma)$. To define the multiplication in $\mathscr{H}$, one imbeds it in the $\mathbb{C}$-vector space $\mathscr{L} = \langle(\Gamma g_i)\rangle$ of formal linear combinations of left cosets, where a basis element $(\Gamma g \Gamma)$ corresponds to the formal sum $\sum_{i=1}^{\mathscr{H}(g)} (\Gamma g_i)$, of the terms in (16). Under this imbedding $\mathscr{H}$ is identified with the subspace $\mathscr{L}^\Gamma$ of elements of $\mathscr{L}$ fixed under the group Γ, which acts on $\mathscr{L}$ according to the rule

$$(\Gamma g)^\gamma = (\Gamma g \gamma) \qquad (g \in \Delta, \ \gamma \in \Gamma).$$

Given any two elements $\sum_i \alpha_i \Gamma(g_i)$, $\sum_j \beta_j (\Gamma g_j) \in \mathscr{L}^\Gamma$, their product is defined to be $\sum_{i,j} \alpha_i \beta_j (\Gamma g_i, g_j) \in \mathscr{L}^\Gamma$; this gives a multiplication on the Hecke algebra which makes $\mathscr{H}$ into an associative algebra, see [3].

In our case let $\Delta = G^+_{\mathbb{Q}}$, $\Gamma = \Gamma_m$. Then the finiteness condition holds, and the Hecke algebra $\mathscr{H}^{(m)} = \mathscr{H}(\Gamma_m, G^+_{\mathbb{Q}})$ is a commutative algebra which decomposes as a tensor product

$$\mathscr{H}^{(m)} = \bigotimes \mathscr{H}_p^{(m)} \qquad (p \text{ prime}),$$

where $\mathscr{H}_p^{(m)}$ is the subalgebra generated by the cosets of the form $(\Gamma g \Gamma)$ with $g \in M_{2m}(\mathbb{Z}[p^{-1}])$, $\mu(g) = p^\delta$. For each prime p the algebra $\mathscr{H}_p^{(m)}$ can be described in a completely explicit way using the Satake isomorphism

$$\mathrm{Sat}\colon \mathscr{H}_p^m \cong \mathbb{C}[X_0^{\pm 1}, X_1^{\pm 1}, \dots, X_m^{\pm 1}]^{W_m}, \tag{17}$$

where the superscript W_m on the right means that we take the subring of elements that are fixed under the action of the Weil group W_m, which acts as the group generated by permutations of the variables $X_1, \dots, X_m$ and also the transformations w_j given by

$$w_j(X_0) = X_0 X_j; \quad w_j(X_j) = X_j^{-1}, \quad w_j(X_i) = X_i \qquad (i \neq j, 0).$$

A basic fact in the theory of Siegel modular forms is that the Hecke algebra $\mathscr{H}^{(m)}$ acts on the finite-dimensional spaces $\mathscr{U}_k^m$ and $\mathscr{S}_k^m$. To define this action, we first give the weight k action of an element $g = \left(\begin{smallmatrix} a & c \\ b & a \end{smallmatrix}\right) \in G_{\mathbb{R}}^+$ on a function $f\colon H_m \to \mathbb{C}$ according to the formula

$$(f|_k g)(z) = \det(cz + d)^{-k} f(g(z)) \det g^{k-(m+1)/2} \tag{18}$$

(this is the notation of Petersson-Andrianov, which gives a convenient way of writing the Hecke operators; we note that with this definition of the action the scalar matrices act nontrivially on functions, although they act trivially as transformations of H_m).

Now let $\Gamma g \Gamma = \bigcup_i \Gamma g_i$ with g, $g_i \in G_{\mathbb{Q}}^+$ be a double coset. We set

$$f|_k(\Gamma g \Gamma) = \sum_i f|_k g_i. \tag{19}$$

Then the definition (19) does not depend on the choice of representatives, and it determines a map $(\Gamma g \Gamma)\colon \mathscr{U}_k^m \to \mathscr{U}_k^m$ which takes $\mathscr{S}_k^m$ to itself. Consequently, we have a representation of the commutative algebra $\mathscr{H}^{(m)}$ in the spaces $\mathscr{U}_k^m$ and $\mathscr{S}_k^m$. It turns out that this representation in $\mathscr{S}_k^m$ is diagonalizable in a certain basis. This fact can be derived by linear algebra from the selfadjointness of the Hecke operators $(\Gamma g \Gamma)$ with respect to the Petersson scalar product in $\mathscr{S}_k^m$; the latter scalar product is defined for $f \in \mathscr{S}_k^m$ and $h \in \mathscr{U}_k^m$ by the formula

$$\langle f, h \rangle = \int_\Phi \overline{f(z)} h(z) \det y^{k-(m+1)}\, dx\, dy,$$

where $\Phi = \Gamma_m \backslash H_m$ is a fundamental domain and $z = x + iy \in H_m$.

Now let f be an eigenfunction for the Hecke algebra. This means that there is a homomorphism $\lambda_f\colon \mathscr{H}^{(m)} \to \mathbb{C}$ such that for all $X \in \mathscr{H}^{(m)}$ we have $f|X = \lambda_f(X) f$. Then, using the Satake isomorphism for all p, the definition of the homomorphism λ_f can be extended to the operators corresponding to the variables X_i in (17); the latter do not act as operators on $\mathcal{M}_k^m$, but rather they correspond to certain matrix operators that take f to a modular form for a congruence subgroup. If we set $\alpha_i(p) = \lambda_f(\mathcal{S}at^{-1}(X_i))$, then it follows from the description of the Weil group that $\alpha_0(p), \alpha_1(p), \dots, \alpha_m(p)$ is a set of nonzero complex numbers which is invariant under the transformations in the group W_m. These numbers are called the p-parameters of the form f; they uniquely determine the homomorphism λ_f.

To give the definition of the zeta-function of a modular form f, we first define the standard element $g_p \in \mathrm{GL}_{2m+1}(\mathbb{C})$ and the spinor element $h_p \in \mathrm{GL}_{2m}(\mathbb{C})$ as the diagonal matrices

$$g_p = \operatorname{diag}\{1, \alpha_1(p), \dots, \alpha_m(p), \dots, \alpha_m(p)^{-1}\}, \tag{20}$$

$$\begin{gathered} h_p = \operatorname{diag}\{\alpha_0(p), \alpha_{i_1}(p), \dots, \alpha_{i_r}(p)\}, \\ 1 \le i_1 < \dots < i_r \le m, \quad 0 \le r \le m. \end{gathered} \tag{21}$$

Then for an arbitrary Dirichlet character $\chi \bmod N$ we can define the spinor zeta-function [1]

$$\mathcal{Z}(s, f, \chi) = \prod_p \mathcal{Z}^{(p)}(s, f, \chi) \tag{22}$$

and the standard zeta-function [4]

$$\mathcal{D}(s, f, \chi) = \prod_p \mathcal{D}^{(p)}(s, f, \chi), \tag{23}$$

where the p-factor is given by means of the characteristic polynomial of h_p and g_p, respectively:

$$\begin{aligned} \mathcal{Z}^{(p)}(s, f, \chi) &= \det(1_{2m} - \chi(p)p^{-s}h_p)^{-1}, \\ \mathcal{D}^{(p)}(s, f, \chi) &= \det(1_{2m+1} - \chi(p)p^{-s}g_p)^{-1}. \end{aligned}$$

The terminology comes from the fact that g_p may be regarded as an element of the orthogonal group SO_{2m+1} in its standard representation in GL_{2m+1} as the group of isometries of the quadratic form with matrix

$$\left(\begin{array}{c|ccc} 1 & 0 & \cdots & 0 \\ \hline 0 & 0_m & \cdots & 1_m \\ 0 & & & \\ \vdots & & & \\ 0 & 1_m & \cdots & 0_m \end{array}\right)$$

and h_p may be regarded as the preimage of g_p under the universal covering of SO_{2m+1} by the spinor group Spin_{2m+1}, realized by means of the spinor representation in GL_{2m}. Here the group Spin_{2m+1} is the Langlands dual of the group $G = G\mathrm{Sp}_m$, and the group SO_{2m+1} is dual to the group Sp_m. This construction of the zeta-function is a special case of the general construction of Langlands automorphic L-functions from automorphic forms on reducible groups over global fields (see [12]).

In connection with these definitions we note that the characteristic roots of the matrices g_p and h_p are permuted under the action of the Weil group W_m, and so, using the Satake isomorphism, the coefficients of the corresponding characteristic polynomials can be expressed in terms of the values of the homomorphism λ_f on $\mathscr{H}_p^{(m)}$. Moreover, one easily sees that all of the roots of the spinor matrix h_p are obtained from a single root $\alpha_0(p)$ by means of a sequence of transformations $w_j \in W_m$. We further note that the definitions (22) and (23) of the spinor and standard zeta-functions can be extended to cusp forms $f \in \mathscr{S}_k^m(C, \psi)$ of weight k for the congruence subgroup $\Gamma_0^m(c)$ with Dirichlet character $\psi \bmod c$ where one assumes that f is an eigenfunction for all of the p-Hecke algebras defined from the subgroup $\Gamma_0^m(c)$ for $p \nmid c$.

§5. Holomorphicity properties and algebraicity of special values of standard zeta-functions

In formulating the results on holomorphicity and on algebraicity of the special values of the standard zeta-functions $\mathscr{D}(s, f, \chi)$ of a Siegel cusp form $f \in \mathscr{S}_k^m(c, \psi)$ of even genus m and weight k for the group $\Gamma_0^m(c)$ with Dirichlet character $\psi \bmod c$ it is convenient to introduce the normalized zeta-functions

$$\mathscr{D}^-(s, f, \chi) = (2\pi)^{-m(s+k-(m+1)/2)} \prod_{j=1}^{m} \Gamma(s+k-j)\mathscr{D}(s, f, \chi), \quad (24)$$

$$\mathscr{D}^+(s, f, \chi) = \frac{2i^\delta \Gamma(s)\cos(\pi(s-\delta)/2)}{(2\pi)^s}\mathscr{D}^-(s, f, \chi), \quad (25)$$

$$\mathscr{D}^*(s, f, \chi) = \pi^{-(s+\delta)/2}\Gamma((s+\delta)/2)\mathscr{D}^-(s, f, \chi), \quad (26)$$

where $\delta = 0$ or 1 according to the relation $\psi\chi(-1) = (-1)^\delta$.

THEOREM (Analytic properties of standard zeta-functions). *Consider a Siegel cusp form*

$$f(z) = \sum_{\xi > 0} a(\xi) e_m(\xi z) \in \mathscr{S}_k^m(c, \psi),$$

and let χ be an arbitrary Dirichlet character modulo $N \geq 1$, not necessarily primitive. Suppose that $k \geq m + \nu$, where $\nu = 0, 1$ and $\chi(-1) = (-1)^\nu$, and suppose that $c\det(2\xi_0)|N$ for some matrix ξ_0 for which $a(\xi_0) \neq 0$. Then the function $\mathscr{D}^*(s, f, \chi)$ extends holomorphically to all $s \in \mathbb{C}$ with the possible exception of a simple pole at $s = 1$ in the case when the character $\chi^2\psi^2$ is trivial.

This theorem is a simple generalization of a result of Andrianov and Kalinin [4]; the book [15] contains a proof based on the detailed study by

Shimura [27] and Feit [17] of the holomorphicity and poles of the analytic continuation of Eisenstein series.

THEOREM (Algebraic properties of special values of standard zeta-functions). *If $k > 2m+2$ and $f \in \mathcal{S}_k^m(c, \psi)$ then in the above notation:*

a) *for integers s with $1 \leq s \leq k - \nu - m$ (where the character $\chi^2\psi^2$ must be nontrivial if $s = 1$) we have*

$$\mathcal{D}^+(s, f, \chi)/\langle f, f\rangle \in K = \mathbb{Q}(f, \lambda_f, \psi, \chi), \tag{27}$$

where $\langle f, f\rangle$ is the Petersson square of the cusp form f, and $K = \mathbb{Q}(f, \lambda_f, \psi, \chi)$ is the field generated by the Fourier coefficients of f, the eigenvalues of the Hecke operators acting on f (i.e., the values of the homomorphism λ_f on the elements $(\Gamma g \Gamma)$), and also the values of the Dirichlet characters ψ and χ;

b) *for integers s with $1 - k + m + \nu \leq s \leq 0$ we have*

$$\mathcal{D}^-(s, f, \chi)/\langle f, f\rangle \in K. \tag{28}$$

This theorem is also proved in [15]; see §7 of Chapter 3. The proof gives some more precise information about the action of the automorphisms $\sigma \in \operatorname{Aut}\mathbb{C}$ on the special values (27) and (28) in terms of the action of σ on f, ψ, χ.

§6. Nonarchimedean standard zeta-functions of Siegel cusp forms

To simplify the statement of our basic result, we make the additional assumption that f is a modular form for the group $\Gamma^m = \mathrm{Sp}_m(\mathbb{Z})$, i.e., $c = 1$, and we also assume that there exists $\xi_0 > 0$ such that $a(\xi_0) = 1$ and $\det(2\xi_0) = 1$. An essential assumption we need is that the cusp form f is p-ordinary: $f : (|i_p(\alpha_0(p))|_p = 1)$ for a fixed imbedding

$$i_p : \overline{\mathbb{Q}} \to \mathbb{C}_p. \tag{29}$$

When verifying this property, the number $\alpha_0(p)$ can be replaced by any of the numbers $\alpha_0(p)\alpha_{i_1}(p)\cdots\alpha_{i_r}(p)$ $(1 \leq i_1 < \cdots < i_r \leq m)$ which make up the 2^m characteristic roots of the p-factor of the spinor zeta-function of the cusp form f.

THEOREM (Nonarchimedean standard zeta-functions). *Suppose that the cusp form $f \in \mathcal{S}_k^m$ is p-ordinary, and let $k > 2m+2$, m even. Then for any integer $c > 1$ not divisible by p there exist bounded $\mathbb{C}_p$-analytic functions*

$$\mathcal{D}^{c+}(x, f),\ \mathcal{D}^{c-}(x, f) : X_p \to \mathbb{C}_p,$$

which are uniquely determined by the following conditions: for all nontrivial Dirichlet characters $\chi \in X_p^{\mathrm{tors}}$

a) *the following equality holds for integers* s *with* $1 \le s \le k - m - \nu$:

$$\mathscr{D}^{c+}(\chi x_p^s, f) = i_p \left[\frac{G_m(\chi) c_\chi^{m(s+k-1-m)}}{\alpha_0(c_\chi)^2} \frac{c_\chi^s}{G(\bar\chi)} (1 - \bar\chi^2(c) c^{-2s}) \frac{\mathscr{D}^+(s, f, \chi)}{\langle f, f \rangle} \right];$$

b) *the following equality holds for integers* s *with* $1 - k + \nu + m \le s \le 0$:

$$D^{c-}(\chi x_p^s, f) = i_p \left[\frac{G_m(\chi) c_\chi^{m(s+k-1-m)}}{\alpha_0(C_x)^2} (1 - \chi^2(c) c^{2s-2}) \frac{\mathscr{D}^-(s, f, \chi)}{\langle f, f \rangle} \right],$$

where

$$G_m(\chi) = \sum_{h \in M_m(\mathbb{Z}) \bmod c_\chi} \chi(\det h) e_m(h/c_\chi)$$

denotes the Gauss sum of genus m *for the primitive Dirichlet character* χ *modulo* $c_\chi = pN_\chi$, $\alpha_0(c_\chi) = \alpha_0(p)^{N_\chi}$ *and we use the convention on Dirichlet characters in* §2.

§7. Proofs of the theorems

The proof of the basic theorem in §6 and also the proofs of the theorems in §5 are based on the connection between the functions $\mathscr{D}(s, f, \chi)$ and Rankin type convolutions. This connection comes from the identity discovered by Andrianov (see [2], [3]):

$$\begin{aligned} &2a(\xi_0) \det \xi_0^{-(s+k-1+\nu)/2} \mathscr{D}(s, f, \chi) \\ &\quad = L_{\mathrm{CM}}(s + (m/2), \psi \chi_{\xi_0} \chi) \\ &\quad \times \prod_{i=0}^{(m/2)-1} L_{\mathrm{CM}}(2s + 2i, \psi^2 \chi^2) L((s - k + 1 - \nu)/2, f, \Theta_{2\xi_0}^{(\nu)}(\chi)), \end{aligned} \tag{30}$$

where $L(s, f, \Theta_{2\xi_0}^{(\nu)}(\chi))$ denotes the convolution of the cusp form f and the theta-series $\Theta_{2\xi_0}^{(\nu)}(\chi)$ with Dirichlet character χ, and χ_{ξ_0} is a certain quadratic character. A detailed explanation of these results is contained in Chapter 3 of [15].

The bounded $\mathbb{C}_p$-analytic functions $\mathscr{D}^{c+}(x, f)$, $\mathscr{D}^{c-}(x, f)$ are constructed as the nonarchimedean Mellin transforms of measures which are, in turn, obtained from certain complex-valued distributions associated with $\mathscr{D}(s, f, \chi)$ and defined directly from the identity (30). The algebraicity of the values of these distributions is proved using a Rankin-Selberg integral representation. Here for the Fourier coefficients of the Siegel-Eisenstein series it is essential to use the holomorphic projection operator and the trace operator in the spaces of real-analytic Siegel modular forms. The resulting distributions on $\mathbb{Z}_p^\times$ with algebraic values become bounded p-adic measures after a certain regularization, which depends on a natural number $c > 1$ not divisible by p, as in the case of the Kubota-Leopoldt p-adic zeta-function. Thus, our method is based on a combination of the technique of Rankin-Selberg and the theory of nonarchimedean integration.

We note that the Rankin-Selberg method has recently been extended to other classical groups [24], [25]. On the other hand, a general theorem has been proved about rationality of the Fourier coefficients of Eisenstein series [18] on symmetric domains which admit a complex structure. This enables one to hope that our method can be applied to a broader class of automorphic L-functions. A different approach to the study of the complex-analytic properties of automorphic L-functions is contained in Shahidi's paper [26]. It would be interesting to try to find a nonarchimedean version of his methods.

REFERENCES

1. A. N. Andrianov, *Euler products corresponding to Siegel modular forms of genus* 2, Uspekhi Mat. Nauk **29** (1974), 44–109; English transl. in Russian Math. Surveys **29** (1974).

2. ———, *Euler products of theta-transformations of Siegel modular forms of genus* n, Mat. Sb. **105** (1978), 291–341; English transl. in Math. USSR-Sb. **34** (1978).

3. ———, *The multiplicative arithmetic of Siegel modular forms*, Uspekhi Mat. Nauk **34** (1979), 67–135; English transl. in Russian Math. Surveys **34** (1979).

4. A. N. Andrianov and V. L. Kalinin, *On the analytic properties of standard zeta functions of Siegel modular forms*, Mat. Sb. **106** (1978), 323–339; English transl. in Math. USSR-Sb. **35** (1979).

5. M. M. Vishik, *Non-Archimedean measures associated with Dirichlet series*, Mat. Sb. **99** (1976), 248–260; English transl. in Math. USSR-Sb. **28** (1976).

6. P. F. Kurchanov, *Local measures connected with Jacquet-Langlands cusp forms over fields of CM-type*, Mat. Sb. (N.S.) **108(150)** (1979), 483–503, 639; English transl. in Math. USSR-Sb. **36** (1980).

7. Yu. I. Manin, *Periods of cusp forms and p-adic Hecke series*, Mat. Sb. **92** (1973), 378–401; English transl. in Math. USSR-Sb. **21** (1973).

8. R. P. Langlands, *Euler Products*, Yale University Press, New Haven, CT, 1971.

9. Yu. I. Manin, *Non-Archimedean integration and p-adic Jacquet-Langlands L-functions*, Uspekhi Mat. Nauk **31** (1976), 5–54; English transl. in Russian Math. Surveys **31** (1976).

10. J. Milnor and J. Stasheff, *Lectures on Characteristic Classes*, Princeton University Press, Princeton, NJ, 1957.

11. A. A. Panchishkin, *Local measures corresponding to Euler products in number fields*, Algebra, English transl., Amer. Math. Soc. Transl. Ser. 2 **137** (1987), MGU, Moscow, 1–13, 1982, pp. 119–138.

12. ———, *Automorphic L-functions and the functoriality principle*, Automorphic forms, representations and L-functions, Mir, Moscow, 1984, pp. 249–286.

13. ———, *Convolutions of Hilbert modular forms and their non-Archimedean analogs*, Mat. Sb. **136** (1988), 574–587; English transl. in Math. USSR-Sb. **64** (1989).

14. ———, *Non-Archimedean Rankin L-functions and their functional equations*, Izv. Akad. Nauk SSSR Ser. Mat. **52** (1988), 336–354; English transl. in Math. USSR-Izv. **32** (1989).

15. ———, *Non-Archimedean automorphic zeta-functions*, MGU, Moscow, 1988.

16. ———, *A functional equation of the non-Archimedean Rankin convolution*, Duke Math. J. **54** (1987), 77–89.

17. P. Feit, *Poles and residues of Eisenstein series for symplectic and unitary groups*, Mem. Amer. Math. Soc. **61**, no. 346 (1986).

18. M. Harris, *The rationality of holomorphic Eisenstein series*, Invent. Math. **63** (1981), 305–310.

19. K. Iwasawa, *Lectures on p-adic L-functions*, Ann. of Math. Stud. **74** (1972).
20. N. M. Katz, *p-adic L-functions for CM-fields*, Invent. Math. **48** (1978), 199–297.
21. T. Kubota and H.-W. Leopoldt, *Eine p-adische Theorie der Zetawerte I. Einführung der p-adischen Dirichletschen L-Funktionen*, J. Reine Angew. Math. **214–215** (1964), 328–339.
22. B. Mazur and H. P. F. Swinnerton-Dyer, *Arithmetic of Weil curves*, Invent. Math. **25** (1974), 1–61.
23. B. Mazur and A. Wiles, *Class fields of Abelian extensions of Q*, Invent. Math. **76** (1984), no. 2, 179–330.
24. I. I. Piatetski-Shapiro and S. Rallis, *L-functions of automorphic forms on simple classical groups*, Modular Forms (Durham 1983), Horwood, Chichester, 1984, pp. 251–261.
25. S. Gelbart, I. I. Piatetski-Shapiro, and S. Rallis, *Explicit constructions of automorphic L-functions*, Lecture Notes in Math., vol. 1254, Springer-Verlag, Berlin and New York, 1987.
26. F. Shahidi, *On the Ramanujan conjecture and finiteness of poles for certain L-functions*, Ann. of Math. **127** (1988), 547–584.
27. G. Shimura, *On Eisenstein series*, Duke Math. J. **50** (1983), 417–476.

Translated by N. KOBLITZ

Amer. Math. Soc. Transl.
(2) Vol. **151**, 1992

Resolution of Singularities in One-Parameter Analytic Families of Differential Equations

UDC 517.925.7+512.761

S. I. TRIFONOV

Introduction

The resolution of singularities plays a major role in the study of differential equations. It allows a wide variety of phase portraits to be reduced to an explicit set of particular cases for which the theory of normal forms is well worked out and can be effectively applied.

Fairly recently (see [7], [1], Ch. 5, §1 and also [6], Appendix 1) a proof has been obtained for the Bendixson–Dumortier theorem, which states that an individual analytic direction field on the real or complex plane may be reduced by a finite number of σ-processes (blowings-up) to an analytic direction field all of whose singularities are elementary. In essence, the theorem says that for the study of planar differential equations the σ-process is in a certain sense a sufficient tool.

A vital next step is to generalize the process of resolution of singularities from individual fields to families of them. In this paper we give such a generalization for one-parameter families (see §2). The construction we propose has two essential features. The first is the necessity of transforming the structure of the parameter space through σ-processes and finite-sheeted coverings: this involves "reproduction" of copies of individual vector fields. The second is that in order to resolve singularities in families we have to carry out σ-processes not only at essential singularities of the family of direction fields (see §1) but also at some points that are not singularities of the family.

The main result in the resolution of singularities in families of differential equations is an analog of the Bendixson–Dumortier theorem: by means of

1991 *Mathematics Subject Classification*. Primary 32S45; Secondary 58F14.

this construction any analytic one-parameter family of direction fields can be reduced to a family all of whose singularities are elementary.

In this paper we state an analog of the Bendixson–Dumortier theorem for the case of a complex analytic one-parameter family of direction fields (§2) and give the main intermediate results on which the proof is based.

Work is currently in progress to generalize the result to complex-analytic families with many parameters and also to real-analytic families.

The author thanks A. Shcherbakov who helped greatly in formulating and proving the Section Theorem in §4, and also Yu. S. Il′yashenko for his continued interest in the work and many helpful comments.

§1. Families of analytic direction fields depending on a parameter

Suppose we are given a complex two-dimensional fibration with complex one-dimensional parameter base space B and three-dimensional total space M; we write this as $(M, B, \pi : M \to B)$. (By a fibration we mean the following: in a small neighborhood of any point of the total space and a corresponding neighborhood of the space B it is possible to define coordinate charts (x, y, ε) and ε, where $x, y \in \mathbf{C}$, $\varepsilon \in \mathbf{C}$, in which the mapping has the form of the standard projection $(x, y, \varepsilon) \mapsto \varepsilon$.)

We denote a typical fiber $\pi^{-1}(\varepsilon)$, $\varepsilon \in B$, by F_ε. Each fiber is a 2-dimensional complex manifold.

In this paper by analytic manifold we mean an analytic manifold having a finite number of connected components. In some cases a manifold N is denoted by a list of its connected components: $N = \{N_i\}$.

DEFINITION.

1. A *complex-analytic direction field* on a complex manifold M is an analytic section with singularities of the projectivized tangent bundle of M, given in a neighborhood of each point of M by an analytic vector field not identically zero.
2. The *singularities* of the direction field are precisely those points of M at which the singular section has an irremovable singularity when projected by the tangent bundle projection onto M.
3. An analytic direction field α on the manifold M that is tangent to the fibers $\{F_\varepsilon\}$ is called an *analytic family* of direction fields.

Throughout this paper we shall adopt the convention that all objects appearing in statements of results are complex-analytic.

LEMMA 1.1. *The set Ω of singularities of an analytic family α of direction fields on a manifold M is an analytic set purely of codimension* 2.

Consider an analytic family α of direction fields. Fix an arbitrary value for the parameter $\varepsilon \in B$. Consider the restriction of α to the fiber F_ε, and the set Ω_ε of singularities that lie in this fiber: $\Omega_\varepsilon = \Omega \cap F_\varepsilon$.

The subset Ω_ε is analytic. Two situations are possible, according to its dimension:

1) Ω_ε is 0-dimensional: all singularities are isolated;
2) Ω_ε is 1-dimensional: there exists a curve of nonisolated singularities.

In case 2) the direction field α_ε can, as a field on the 2-dimensional manifold F_ε, be expanded uniquely to all the regular points of Ω_ε. Therefore α_ε may also as in case 1) be regarded as a field with isolated singularities.

DEFINITION. The points of the total manifold M that are singularities of the restriction α_ε of the total field α to the fiber F_ε, when α_ε is expanded over the 2-dimensional fiber to give a field with isolated singularities, are called *essential singularities.* The set of essential singularities will be denoted by Ω_{ess}.

REMARK. The intersection of Ω_{ess} with an arbitrary fiber F_ε is a discrete set.

EXAMPLE. The codimension of Ω_{ess} may be greater than two. For example, in the family of fields

$$\begin{cases} \dot{x} = xy + \varepsilon, \\ \dot{y} = y^2, \qquad x, y, \varepsilon \in \mathbf{C}, \end{cases}$$

the set of essential singularities has codimension 3 since it consists of a single point: $\Omega_{\text{ess}} = \{(0, 0, 0)\}$.

DEFINITION. Consider a purely 1-dimensional analytic set Ψ contained in the total space M of the fibration $(M, B, \pi : M \to B)$. The subset Ψ_{reg} of *regular points* is the union of all those points of Ψ that have a neighborhood that is mapped biholomorphically by the projection into the base B.

LEMMA 1.2. *The closure $\overline{\Omega}_{\text{ess}}$ of Ω_{ess} is an analytic set of codimension not greater than two. There is a discrete set D such that any essential singularity $q \in \Omega_{\text{ess}}$ not belonging to D is a regular point of a purely 1-dimensional irreducible component of Ω_{ess}.*

LEMMA 1.3. *For any point $p \in M$ there is a neighborhood $W_p \subset M$ of p and a purely 1-dimensional analytic set $\Psi_p \subset W_p$ such that any essential singularity $q \in \Omega_{\text{ess}} \cap W_p$ is a regular point of some irreducible component of Ψ_p.*

In the example above it is easy to see that Ψ_p is uniquely defined. For example, we may take it to be the set $\{(x, y, \varepsilon) | x = y = 0, \varepsilon \in \mathbf{C}\}$.

COROLLARY 1.1. *For any compact set K in the total space there exists an open neighborhood $L \subset M$, a finite open cover $\{U_i\}$ of $\pi(K) \subset B$ and a purely 1-dimensional set Ψ_i in $\pi^{-1}(U_i) \cap L$ such that any essential singularity lying in the open subset $\pi^{-1}(U_i) \cap L$ of M is a regular point of some irreducible component of Ψ_i. Here $\pi(L) = \bigcup_i U_i$.*

§2. The definition of a blown-up family

DEFINITION. If B is a 1-dimensional analytic manifold, then an analytic (not necessarily connected) 1-dimensional manifold $\tilde{B}$ with an analytic mapping $\rho : \tilde{B} \to B$ is called an *expansion* ([1]) of B if

1. there is a discrete subset $A \subset B$ such that the restriction of ρ to $B \backslash A$ is a local analytic diffeomorphism;
2. for any $\varepsilon \in B$ there exists a neighborhood $U_\varepsilon \subset B$ and a finite collection of open subsets $\{\tilde{U}_{\varepsilon,i}\} \subset \tilde{B}$ such that ρ restricted to the union of the $\tilde{U}_{\varepsilon,i}$ maps surjectively onto U_ε: $\rho(\bigcup_i \tilde{U}_{\varepsilon,i}) = U_\varepsilon$.

EXAMPLES OF EXPANSIONS.

1) $\tilde{B} = \{U_i\}$, a finite open cover of the base B, with $\rho|_{U_i} = \mathrm{id}|_{U_i}$;

2) $\rho : \tilde{B} \to B$, a finite-sheeted branched covering.

The expansions that we shall be applying in what follows will be finite superpositions of finite open covers and finite-sheeted coverings.

We turn now to the definition of a blown-up family.

DEFINITION. We say that we have a *blown-up fibration* for the fibration $(M, B, \pi : M \to B)$ if we have the following:

- a new fibration $(\tilde{M}, \tilde{B}, \tilde{\pi} : \tilde{M} \to \tilde{B})$;
- an expansion of the base $\rho : \tilde{B} \to B$;
- a holomorphic mapping $H : \tilde{M} \to M$ of the total manifolds

such that the diagram

$$\begin{array}{ccc} \tilde{M} & \xrightarrow{H} & M \\ {\scriptstyle\tilde{\pi}}\downarrow & & \downarrow{\scriptstyle\pi} \\ \tilde{B} & \xrightarrow{\rho} & B \end{array}$$

commutes. Here the restriction of H to any fiber $\tilde{F}_{\tilde{\varepsilon}}$ is a holomorphic map $\tilde{F}_{\tilde{\varepsilon}} \to F_{\rho(\tilde{\varepsilon})}$ and may be expressed as the composition of a finite number of reverse σ-processes.

DEFINITION. We have a *blown-up family* of direction fields $(\tilde{M}, \tilde{B}, \tilde{\pi} : \tilde{M} \to \tilde{B}, \tilde{\alpha})$ if the fibration $(\tilde{M}, \tilde{B}, \tilde{\pi} : \tilde{M} \to \tilde{B})$ is a blow-up for the fibration $(M, B, \pi : M \to B)$, and the holomorphic family $\tilde{\alpha}$ of direction fields on the fibers $\tilde{F}_{\tilde{\varepsilon}}$ is taken by the mapping H of the total spaces into the family α on $F_{\rho(\tilde{\varepsilon})}$: $H_* \alpha = \tilde{\alpha}$.

DEFINITION. A blown-up family of direction fields $(\tilde{M}, \tilde{B}, \tilde{\pi} : \tilde{M} \to \tilde{B}, \tilde{\alpha}, \rho, H)$ is a *good blow-up* if all the essential singularities of $\tilde{\alpha}$ restricted to each respective fiber are elementary (see §4).

DEFINITION. The restriction of the fibration $(M, B, \pi : M \to B)$ to an open subset $L \subset M$ is written as $(L, \pi(L), \pi|L)$.

MAIN THEOREM. *The restriction of a complex analytic family of direction fields to any open precompact set in the total space has a good blow-up.*

([1]) Editor's note. The Russian term is "развертка" (literally "development").

§3. Expansions of the base and statement of the Section Theorem

In this section we describe a geometrical construction for an expansion of the base which enables us to obtain blown-up families with simpler configurations of essential singularities. We then formulate the Section Problem. All the statements and results in this section are given in geometric terms without recourse to terminology from the theory of differential equations.

LEMMA 3.1. *Given a fibration* $(M, B, \pi : M \to B)$ *and a holomorphic surjective mapping* $\tau : \tilde{B} \to B$ *of an analytic* 1*-dimensional manifold* $\tilde{B}$ *onto* B, *there is a well-defined fibration* $(\tilde{M}, \tilde{B}, \tilde{\pi} : \tilde{M} \to \tilde{B})$ *with new base* $\tilde{B}$ *and an analytic mapping* $h : \tilde{M} \to M$ *of total manifolds such that the following hold*:

1) *The diagram*

$$\begin{array}{ccc} \tilde{M} & \xrightarrow{h} & M \\ \tilde{\pi}\downarrow & & \downarrow\pi \\ \tilde{B} & \xrightarrow{\tau} & B \end{array}$$

commutes.

2) *The restriction of* h *to any fiber* $\tilde{F}_{\tilde{\varepsilon}}$, $\tilde{\varepsilon} \in \tilde{B}$, *gives an analytic diffeomorphism between the fibers* $\tilde{F}_{\tilde{\varepsilon}}$ *and* $F_{\tau(\tilde{\varepsilon})}$.

DEFINITION. A fibration defined as above is called a *fibration with expanded base* if the map τ is an expansion (see §2).

If we are given a family of direction fields on the fibration then we may suppose that after expanding the base the copies of the fibers are reproduced together with the direction fields defined on them. We call the resulting family a *family with expanded base.*

LEMMA 3.2 (Selection of sections). *With notation as before, for any analytic mapping* $\phi : \tilde{B} \to M$ *such that* $\pi \circ \phi = \tau$ *there exists a mapping* $\tilde{\phi} : \tilde{B} \to \tilde{M}$ *for which* $\tilde{\pi} \circ \tilde{\phi} = \mathrm{id}$, $h \circ \tilde{\phi} = \phi$. *Hence in the new total space at least one section can be found contained in the set* $h^{-1}(\phi(\tilde{B}))$.

LEMMA 3.3 (Behavior of sections under expansion of the base). *On expansion of the base we have*:

1) *the multiplicity of the covering of any analytic set of codimension* 2 *(defined at the regular points) does not change*;
2) *the inverse image of any section remains a section.*

EXAMPLE. Let $\{U_i\}$ be an open cover of the base B. Regard this cover as an expansion with canonical map $\tau : \{U_i\} \hookrightarrow B$. After expansion of the base, the new total space becomes an open cover of the original total space $\{M_i\} \hookrightarrow M : \{h_i\} = \{\mathrm{id}|_{M_i}\}$, $\tilde{\pi}_i = \pi|_{M_i}$.

THEOREM (Section Theorem). *Given a complex analytic fibration* $(M, B, \pi : M \to B)$, *a purely* 1*-dimensional analytic set* $\Psi \subset M$ *and a compact set* $K \subset M$, *there exists an open neighborhood* L *of* K *and an expansion of the manifold* $\pi(L)$ *such that, after expanding the base of the fibration restricted to* L, *the closure of the inverse image of* $\Psi_{\text{reg}} \cap L$ *breaks up into a finite number of sections of connected components of the new fibration.*

LEMMA 3.4. *Let* N *be a precompact open subset of the total space* M *of the fibration* $(M, B, \pi : M \to B)$. *After any expansion of the base is carried out, there is a precompact open set* $\tilde{N}$ *in the new total space with the following properties:*

1) *the image of* $\tilde{N}$ *is* N;
2) *the intersection of* $\tilde{N}$ *with any fiber* $\tilde{F}_{\tilde{\varepsilon}}$, $\tilde{\varepsilon} \in \tilde{B}$, *is either empty or coincides with* $N \cap F_{\rho(\tilde{\varepsilon})}$ *in the original space (in the sense of the biholomorphism between fibers as defined in Lemma* 3.1).

Thus the restriction of the new fibration to $\tilde{N}$ is a fibration with expanded base relative to the original fibration restricted to N.

We shall assume from now on that whenever a precompact open set N is given in the original fibration, it is expanded to $\tilde{N}$ when the base is expanded.

§4. The tools of the Σ-process

We regard the original family as trivially blown-up: $(M, B, \pi : M \to B)$, $\rho = \text{id}$, $H = \text{id}$.

We now present two methods for obtaining blown-up families. Carrying them out in succession leads, as we wish to show, to a construction for a good blow-up.

To describe these methods we suppose that we have a blown-up family $(M, B, \pi : M \to B)$, H, ρ and we obtain from it a "more" blown-up family $(\tilde{M}, \tilde{B}, \tilde{\pi} : \tilde{M} \to \tilde{B})$, $\tilde{H}$, $\tilde{\rho}$.

4.1. Expansion of the base.

LEMMA 4.1. *After expansion of the base of a family of direction fields, the set of essential singularities of the covered family is contained in the inverse image of the set of essential singularities of the original family.*

Let N be a precompact open set in the original fibration $(M, B, \pi : M \to B)$. We restrict the direction field to an open neighborhood L of the compact set $\overline{N}$ to which we shall apply Corollary 1.1. Expand the base $\pi(L)$ by means of a finite partition $\{U_i\}$ defined on it; then N expands to an open precompact set $\tilde{N}$ by Lemma 3.4. By Corollary 1.1, in each of the resulting components M_i there is a purely one-dimensional set Ψ_i with the property that any essential singularities of the family with expanded base $N \cap M_i$ are regular points of the irreducible components of the set Ψ_i.

COROLLARY 4.1. *After any expansion of the base of the family of direction fields obtained above, the set of those essential singularities of the covered family that lie in the corresponding expanded open precompact set is contained in the union of the closures of the inverse images of the subsets of regular points of the* $\{\Psi_i\}$.

COROLLARY 4.2. *Assuming the Section Theorem is proved, for any analytic family of direction fields and open precompact set* N *of the total manifold* M *there exists a neighborhood* L *of the closure of* N *and an expansion* $\tau : \tilde{B} \to \pi(L)$ *of the base such that, when this is carried out over the restriction of the original fibration to* L, *the set* Ω_{ess} *of all essential singularities of the resulting family becomes a subset of the union of a finite number of sections.*

It is easy to see that on expanding the base in the family $(M, B, \pi : M \to B), H, \rho$ we obtain a further blow-up of the original family: $(\tilde{M}, \tilde{B}, \tilde{\pi} : \tilde{M} \to \tilde{B}), \tilde{H}, \tilde{\rho}, \tilde{\alpha} = h_*\alpha, \; \tilde{H} = h \circ H, \; \tilde{\rho} = \tau \circ \rho$.

4.2. Blown-up sections. Given a section of a connected component of the total space, the operation of the σ-process is well defined in all fibers at the points of this section. Hence we obtain a 3-dimensional manifold from this connected component while all the other components remain unchanged. The new total manifold projects onto the same base: $\tilde{\rho} = \rho$.

When a section is blown up there is a qualitative change in the singularity set of the family of direction fields: the singularities that lie on the section disappear, and new singularities are created on the "projective cylinder" that is obtained. If a point of the section is not an essential singularity for the original family then on blowing up the section one saddle singularity appears (see Lemma 5.3). As a result of the σ-process we again obtain an analytic direction field.

LEMMA 4.2. *After blowing up a section of a* 1*-dimensional complex-analytic fibration, the nonsingular inverse image of any other section of the original fibration becomes a section of the blown-up fibration.*

LEMMA 4.3. *After blowing up a section, the full inverse image of a precompact open set remains a precompact open set.*

§5. Blow-ups of points of an individual field

We pass now to the proof of the main theorem of the paper. For this we need to state some results about individual direction fields in the plane.

All the results in this section are stated for individual analytic direction fields on 2-dimensional complex and real manifolds.

5.1. An account of the proof of the Bendixson-Dumortier Theorem. In this subsection we describe the proof of the Bendixson-Dumortier Theorem due to van den Essen (see [6, Appendix 1]).

DEFINITION. The *multiplicity* of a singularity of a direction field on a 2-manifold is the minimum multiplicity of analytic vector fields that give rise to the given direction field.

Since the multiplicity of an isolated singularity of a vector field is finite, the multiplicity of an essential singularity of a direction field that is not completely degenerate is also finite.

DEFINITION. A singularity of a vector field on a 2-manifold is said to be *elementary* if at least one eigenvalue of its linearization is nonzero, and is of "*block*" type if its linearization is a nilpotent Jordan block.

DEFINITION. A singularity of a direction field on a 2-manifold is *elementary* (of "*block*" type) if the germ of the direction field at the singularity can be given by the germ of an elementary singularity (point of "block" type) of a vector field.

PROPOSITION. *A singularity of a vector field that has multiplicity* 1 *is elementary.*

LEMMA 5.1. *When a σ-process is carried out at an isolated singularity of a holomorphic direction field in a region of the complex plane, there is a finite number of singularities on the attached projective line as follows*:

a) *if the original singularity is not elementary or of "block" type then the sum of the multiplicities of the newly-obtained singularities is less than the multiplicity of the original one*;
b) *if the original singularity is elementary then the new singularities are elementary.*

LEMMA 5.2. *Points of "block" type break up into elementary points after a finite number of σ-processes.*

THEOREM (Bendixson-Dumortier). *After a finite number of σ-processes an arbitrary holomorphic direction field with isolated singularities becomes a vector field with only elementary singularities. In other words, the field has a good blow-up.*

PROOF. Denote by μ the sum of the multiplicities of all the singularities of the vector field.

a) Define a *transformation step* to mean the carrying-out of σ-processes at all singularities of the direction field.
b) Carry out μ steps. By Lemma 5.1 all the singularities have then broken up into elementary ones, those of "block" type and those generated from the latter by σ-processes. After each step there is at most a finite number of "block" type points. By Lemma 5.2, the number of σ-processes needed to break up all the "block" type points is bounded above by some natural number ν.
c) Carry out ν steps. After this all the singularities are elementary. The theorem is proved.

5.2. Additional results. In the sequel we shall need the following two lemmas, which can be verified by direct calculation.

LEMMA 5.3. *The result of a σ-process at a nonsingular point is to create one elementary singularity of nondegenerate saddle type.*

DEFINITION. The *degree of degeneracy* of an arbitrary singularity of an analytic direction field is defined to be the minimal number of steps, as in the proof of the Bendixson-Dumortier Theorem, that are needed in order to break up the singularity into elementary ones.

LEMMA 5.4. *The degree of degeneracy of an arbitrary singularity is less than or equal to $\mu + 5$ where μ is the multiplicity of the singularity.*

§6. Reduction of the main theorem to the Section Theorem

Modulo the Section Theorem, the proof repeats the proof given in §5.

The difference lies in the fact that some singularities of separate fibers will be blown up more often than is necessary for blowing up an individual field. In cases when the set of essential singularities contains isolated points, many nonsingular points are blown up as well as the essential singularities. The transformations that we shall carry out are those described in §4: namely, expansion of the base and blow-up of sections.

The key point in the proof is the uniform bound on the total multiplicity of all the essential singularities in the fibers.

LEMMA 6.1. *In any analytic family of direction fields, restricted to an arbitrary precompact open set in the total space, the total multiplicity of all the essential singularities in the fiber is bounded uniformly over the base manifold.*

This lemma is a consequence of a theorem of Gabrielov [2].

We now describe a procedure for obtaining a good blow-up of the family.

a) *Description of the step.* Consider a fibration $(M, B, \pi : M \to B)$ and a family α of direction fields on it. If we assume the Section Theorem (see §3) then by Corollary 4.2 there exists a restriction of the fibration to some neighborhood of the closure of a precompact set N and a corresponding expansion of the base such that for the new fibration the set Ω_{ess} of all essential singularities of the family $\tilde{\alpha}$ of direction fields is a subset of the union of a finite number of sections. We now blow up all these sections one by one.

Since, on expanding the base and blowing up sections, those sections that are not the ones being blown up remain sections (Lemmas 3.3 and 4.2), after a finite number of coverings of the base and blow-ups of sections we shall have blown up all sections.

On completing this step we obtain a new family of direction fields in which all essential singularites of the original family that are in N have been blown up at least once each. By Lemmas 3.4 and 4.3 in the new fibration there will be a new precompact open set $\tilde{N}$ such that the restriction of the new fibration to $\tilde{N}$ is a blow-up of the restriction of the old fibration to N.

b) We carry out $\mu + 5$ steps, where μ is the maximum sum of the multiplicities of singularities in the fiber (which exists by 6.1). After this, by Lemma 5.4, all the essential singularities of the family which originally lay in N have become elementary.

The proof is now complete, modulo the Section Theorem.

§7. Proof of the Section Theorem

The proof is based on the theory of Riemann functions ([4]).

LEMMA 7.1. *Let Ψ be a purely 1-dimensional irreducible set in a neighborhood of zero in $\mathbf{C}^3_{x,y,\varepsilon}$, with projection $\pi : \mathbf{C}^3_{x,y,\varepsilon} \to \mathbf{C}_\varepsilon$, $(x, y, \varepsilon) \mapsto \varepsilon$. Then either*

1) *the set Ψ contains no regular points for the projection π, or*
2) *there is a small neighborhood of zero such that the set $\Psi \cap W$ is discrete in the fibers: for any $\varepsilon \in \mathbf{C}$ the set $\Psi \cap W \cap F_\varepsilon$ is discrete. In this case the map $\pi|_{\Psi\cap W}$ is proper, and hence ([5]) gives a finite-sheeted covering $\pi|_{\Psi\cap W} : \Psi \cap W \to \mathbf{C}_\varepsilon$ of a full neighborhood of zero, unramified away from the point $\varepsilon = 0$.*

LEMMA 7.2. *Let Ψ be a purely 1-dimensional set in the fibration $(M, B, \pi : M \to B)$. Then any point $p \in M$ has a neighborhood $W \subset M$ such that either*

1) *the set $\overline{\Psi}_{\mathrm{reg}} \cap W$ is the union of a finite number of sections, or*
2) *there exists an expansion $\tau : \tilde{B} \to \pi(W)$ which, when applied to a fiber restricted to W, contains in the closure of the inverse image of the set Ψ_{reg} of regular points more sections than there are in $\overline{\Psi}_{\mathrm{reg}} \cap W$ for the original fibration restricted to W.*

PROOF. Fix a small neighborhood W of p. Consider an irreducible component Ψ_1 of the purely one-dimensional set Ψ, and suppose Ψ_1 is not a section. If Ψ_1 contains no regular points it is not relevant to the statement of the lemma and may be disregarded. Otherwise, by Lemma 7.1, we may restrict W so that $\Psi_1 \cap W$ is a finite-sheeted covering of $\pi(W)$ unramified everywhere except at p.

The set $\Psi_1 \cap W$ is therefore a Riemann surface and may be continued to a Riemann surface $\overline{\Psi}_1$ which is a finite-sheeted branched covering of $\pi(W)$ ([4]). The multiplicity of the covering $\overline{\Psi}_1$ must be greater than one, otherwise Ψ_1 is a section of the fibration. Take $\tilde{B}$ to be $\overline{\Psi}_1$ and τ to be its covering map onto the base $\pi(W)$. By Lemmas 3.3 and 3.4, the inverse image of Ψ_1 contains at least one section, and all the previously-existing sections remain sections. The lemma is proved.

PROOF OF THE SECTION THEOREM. Suppose Ψ is not a union of sections. Since the set K in the statement of the theorem is compact there exists a finite open cover of K by neighborhoods $\{W_i\}$ each of which satisfies the statement of Lemma 7.2. Let $L = \bigcup_i W_i$. Restrict the fibration to L, and expand the base $\pi(L)$ using the finite open cover $\{\pi(W_i)\}$. If in some component M_i the set $\Psi \cap M_i$ is not the union of a finite number of sections, we carry out in M_i an expansion τ_i of the base, as defined in Lemma 7.2. As a result of this, the multiplicity of the covering of the closure of the inverse image of Ψ_{reg} remains equal to the multiplicity of the covering of Ψ_{reg}, and the number of sheets that correspond to sections increases in those components where it was less than the multiplicity of the covering of the whole set.

Consider a precompact open neighborhood N of K in L: $K \Subset N \Subset L$. By Lemma 3.4 the resulting fibration contains an expanded precompact set $\tilde{N}$. If not every component of the closure of the inverse image of Ψ_{reg} decomposes into sections, we carry out the above procedure once again, taking the closure of N as the compact set in the total space of the fibration obtained so far. The set K is compact, the maximum multiplicity of the covering $\pi : \Psi \cap K \to B$ defined on the regular points is finite, and therefore after a finite sequence of processes as above we obtain that in each connected component of the fibration the closure of the inverse image of Ψ_{reg} is a collection of sections of connected components.

If we regard the image in M of the total space so obtained as being the neighborhood L of the compact set K in the original fibration we obtain the assertion of the Section Theorem.

BIBLIOGRAPHY

1. V. I. Arnol′d and Yu. S. Il′yashenko, *Ordinary differential equations*, Itogi Nauki i Tekhniki: Sovremennye Problemy Mat., Fundamental′nye Napravleniya, vol. 1, VINITI, Moscow, 1985, pp. 7–149; English transl. in Encyclopedia of Math. Sci., vol. 1 (Dynamical Systems, I), Springer-Verlag, 1987.

2. A. M. Gabrielov, *Projections of semianalytic sets*, Funktsional. Anal. i Prilozhen. **2** (1968), no. 4, 18–30; English transl. in Functional Anal. Appl. **2** (1968), no. 4.

3. P. A. Griffiths and J. D. Harris, *Principles of algebraic geometry*, Pure and Applied Mathematics, Wiley-Interscience, New York, 1978.

4. O. Forster, *Riemann Surfaces*, Springer-Verlag, Berlin and New York, 1977.

5. E. M. Chirka, *Complex analytic sets*, "Nauka", Moscow, 1985; English transl. Reidel, Dordrecht, Boston (to appear).

6. J. D. Mattei and R. Moussu, *Holonomy and first integrals*, Ann. Sci. Ecole Norm. Sup. (4) **13** (1980), no. 4, 469–523.

7. A. Seidenberg, *Reduction of singularities of the differential equation $A\,dy = B\,dx$*, Amer. J. Math. **90** (1968), 248–269.

Translated by D. R. CHILLINGWORTH

Recent Titles in This Series

(Continued from the front of this publication)

(See the AMS catalog for earlier titles)